# HOW TO MAKE MAPLE SYRUP

From Gathering Sap
to Marketing Your Own Syrup

Alison and Steven Anderson

Storey Publishing

*The mission of Storey Publishing is to serve our customers by publishing practical information that encourages personal independence in harmony with the environment.*

Edited by Sarah Guare and Deborah Burns
Art direction by Cynthia N. McFarland, based on a book design by Alethea Morrison
Text production by Jennifer Jepson Smith

Cover illustration by © Lisel Ashlock
Interior illustrations by © Elayne Sears
Maps by Ilona Sherratt
Photograph on page v courtesy of the authors

Indexed by Christine R. Lindemer, Boston Road Communications

© 2014 by Steven and Alison Anderson

All rights reserved. No part of this book may be reproduced without written permission from the publisher, except by a reviewer who may quote brief passages or reproduce illustrations in a review with appropriate credits; nor may any part of this book be reproduced, stored in a retrieval system, or transmitted in any form or by any means — electronic, mechanical, photocopying, recording, or other — without written permission from the publisher.

The information in this book is true and complete to the best of our knowledge. All recommendations are made without guarantee on the part of the author or Storey Publishing. The author and publisher disclaim any liability in connection with the use of this information.

Storey books are available for special premium and promotional uses and for customized editions. For further information, please call 1-800-793-9396.

**Storey Publishing**
210 MASS MoCA Way
North Adams, MA 01247
*www.storey.com*

Printed in the United States by McNaughton & Gunn, Inc.
10 9 8 7 6 5 4 3 2 1

Storey Publishing is committed to making environmentally responsible manufacturing decisions. This book was printed on paper made from sustainably harvested fiber.

Library of Congress Cataloging-in-Publication Data on file

# CONTENTS

*For Dad. Thank you for passing on your knowledge*
*of the syrup industry and instilling in me your tireless work ethic.*
*— S.A.*

*For Mom and Dad. — A.K.A.*


## ACKNOWLEDGMENTS

Thank you to Sarah Dorison and Brenda Buck for reading and editing our manuscript, and to Deb Burns and Sarah Guare, of Storey Publishing, for their patience and encouragement.

# PREFACE

Anderson's Maple Syrup, located near Cumberland, Wisconsin, has been a family-run business since 1928. There are about 2,500 sugar maples in our family sugar bush, plus another 2,000 maple trees just 5 miles down the road.

## Our Roots

Steve's family purchased the homestead in 1928, and as for many Wisconsin families, the main source of income was dairy farming. Paul Anderson, Steve's grandfather, inspired by childhood memories of sugar making with his father, began tapping trees and making just enough maple syrup for his family and neighbors to enjoy. He collected sap with metal buckets and spouts, and a horse and wagon carried it to the sugarhouse.

Paul Anderson cooks syrup on two 6-by-16-foot Leader Evaporators.

There he boiled it to perfection on an open pan over a handmade stone and cement arch. Paul purchased the first evaporator for the family in 1947 for $680 — a brand-new 5-foot by 16-foot King. As he sold more syrup, he was able to rely less and less on farming for income. In 1957 he sold the last of the dairy cows and turned to syrup making and syrup equipment sales full-time.

Paul's son, Norman, grew up helping his father care for the sugar bush and learning how to produce sweet maple syrup, and eventually Paul and Norman were running the business together. They began selling syrup-making equipment in 1954, when the family became a Leader Evaporator dealer. Anderson's is still a Leader dealer today.

Originally, equipment was delivered from Leader's Vermont headquarters to the North Woods of Wisconsin by train. The train came through a nearby town and parked on a railroad siding, and Paul and Norman unloaded the train car there. Today semi-truck loads of equipment and pallets of syrup go directly in and out of the warehouse, located across the road from the sugarhouse where Paul made syrup.

## Peak Production

In the 1970s we were at peak production, tapping almost 18,000 trees with buckets on various pieces of land across Wisconsin's North Woods. Norman ran the business at this capacity for about 10 years while he expanded the company and increased syrup and equipment sales. Norman hired the company's first full-time employee in 1974.

By 1980 we were able to sell far more syrup than we could produce ourselves. Norman began purchasing syrup from other local producers, who then bought syrup-making equipment from Norman. Anderson's Maple Syrup still operates on this cooperative model. Norman's son Steve sells equipment and buys syrup from many of the same producers his father did, as well as many others.

With fewer than 10 employees, Anderson's is still a small, family-run business, but the company now has a national, even global, presence. Steve has increased syrup sales and taken the Anderson label from a regional family staple to a national syrup brand. We ship syrup to customers as far as Asia. We also have an extensive syrup equipment business with a showroom and a website.

Steve has added technological advancements to the business by utilizing some automated equipment for packaging and by modernizing the way sap is collected in the sugar bush. These days you won't find a single bucket hanging in the woods in the Anderson's sugar bush. We use a modern, tree-friendly tubing system with vacuum, which moves the sap directly to a collection site. From there the sap is transported to the sugarhouse for cooking and bottling.

Our family is proud to participate in Wisconsin's Forest Crop Law program and Conservation Reserve Program. These are both voluntary efforts for agricultural landowners designed to protect natural habitats and preserve them for future generations.

## *Conservation Programs*

The **Forest Crop Law (FCL)** program, part of the Wisconsin Department of Natural Resources, encourages long-term sustainable management of private woodlands. In exchange for participating in FCL, landowners pay reduced property taxes.

The **Conservation Reserve Program (CRP)** is a voluntary USDA and Farm Service Agency (FSA) based program. This program, available to agricultural landowners, encourages members to use portions of their land to further conservation efforts. Participants in CRP are provided with yearly compensation.

## Our Guide for You

We created this guide as a quick reference for beginning syrup makers. We want to encourage those interested in syrup making to tap their maples, so that the legacy and art of this rustic tradition will continue well into the future.

We've included what we believe to be the easiest and most straightforward methods for harvesting and cooking sap and for filtering and bottling finished syrup. There are many variations in the way you can gather sap and make syrup, and innovations are constantly improving the maple industry, but the information in this guide is a good place to start.

We hope you find this guide helpful and that you have continued success in syrup making!

CHAPTER 1

# THE HISTORY OF MAPLE SYRUP

The maple syrup industry has evolved greatly since Native Americans discovered how to make pure maple syrup. Vast improvements have been made in sap collecting and cooking methods and syrup-making equipment. The basic process remains the same, however: Boil sap to evaporate water and make sweet, pure maple syrup.

## NATIVE AMERICANS DISCOVER MAPLE SYRUP

It is generally agreed that Native North Americans were the first to make maple syrup. Various legends describe how maple syrup was discovered:

**The "sap-sicle."** The branch of the maple tree broke during the late winter and sap flowed out of it. The dripping sap froze into an icicle. The repeated thawing and freezing of the sap-sickle concentrated the sugars and made the icicle unusually sweet. A Native American tasted the icicle. He or she noticed that the liquid had a distinct sweetness to it and deduced that it had come from inside the tree, rather than from melting snow.

**Hatchet job.** Another legend involves a Native American hunter who, when he arrived home from an unsuccessful hunt, threw his hatchet at a maple tree. The hatchet stuck in the tree and created a wound. The sap that flowed from the wound in the tree collected in a cooking container that the hunter's wife had placed on the ground beneath the tree. The next day the hunter's wife mistakenly thought her husband had filled her container with water and she began to cook with it. When they tasted their meal, they discovered that it was sweet! After examining the tree and discovering the sap flowing from it, they realized what had happened.

**Substitution.** A woman was cooking supper for her husband. When she became distracted by her quillwork, her boiling pot went dry. She did not have time to melt snow so she poured in some maple sap that she had collected for drinking. Their meal had a delicious sweetness to it. Her husband enjoyed the meal so much, it is said that he licked the pot clean!

## Collecting and Concentrating

It is believed that Native Americans collected sap by slashing the bark of a maple tree and inserting a hollow twig or scraped piece of bark to direct the flowing sap into a wood or bark container sitting on the ground. Native Americans placed hot stones in the sap-filled wooden containers. The heat from the stones prompted evaporation. As the process was repeated over and over, syrup was the eventual product.

Another method Native Americans used to concentrate the sugars in sap was to gather sap and let it freeze. The sugars in the sap did not freeze. When the ice was discarded, a sweet liquid remained, though it was not as flavorful as maple syrup. It takes heat to intensify the natural maple flavor that can be found only in pure maple syrup.

## NATIVE AMERICANS TEACH EUROPEAN SETTLERS

EUROPEAN SETTLERS TRADED GOODS with Native Americans for maple sugar, and they eventually copied Native American practices to make their own. Settlers modified the practices of Native Americans by making smaller holes in the trees and inserting spouts made from the hollowed stems of sumac. Sap was collected in wooden and clay buckets; we display some very early clay sap buckets in our store showroom. Eventually wooden spiles and wooden buckets were replaced by metal or plastic, thus moving syrup making into the modern era.

In 1764 the Parliament of Great Britain passed the Sugar Act, imposing high tariffs on cane sugar imported to the colonies. Maple sweetener, which many colonists could make for themselves, became very popular because of its low cost and availability. There was also some moral opposition to buying and using cane sugar because it was largely produced by slave labor in the British West Indies. During this time, maple sap was primarily concentrated into sugar, which was very popular.

Around the time of the Civil War, sugarcane crops were established in the United States, and it was not necessary to import as much sugar. The Sugar Act was eventually repealed, and cane sugar replaced maple sugar as the dominant sweetener in the United States. The maple producers shifted their focus from sugar making to syrup making and began to promote the use of pure maple syrup.

## MODERNIZING SYRUP MAKING

EARLY SYRUP MAKERS TRANSPORTED SAP in the wooden buckets it was collected in, using shoulder yokes to carry them from the trees to the cook site. As sugar bushes grew in size, it became more difficult to transport sap by hand. Syrup producers hauled sap in wooden tubs, mounted on sleds or wagons and pulled by horses or oxen. Eventually wooden tubs were replaced with metal ones and horses were replaced with tractors. The Andersons' sugar bush utilized a wagon and horses until the early 1960s.

Even prior to World War I, producers attempted to collect sap more efficiently than the one-bucket-per-spout method. They used metal tubes, similar to the way sap is collected via modern plastic tubing systems. Although the metal tubing systems proved to be troublesome — leaky, hard to clean, and somewhat difficult to install — they paved the way for modern plastic tubing.

To produce syrup, some Native Americans boiled sap in clay pans over open fires. Settlers copied this practice by boiling sap in large metal cauldrons that hung on tripods over open fires. Syrup made in these large cauldrons had a very strong flavor and dark color due to the length of time the sap was exposed to heat. Settlers found that finishing the syrup in smaller batches in multiple cauldrons resulted in lighter-colored syrup with a more desirable and even delicate flavor. Producers today still finish syrup in separate pans.

Producers found that cooking sap in flat-bottomed pans increased the rate of evaporation. By the mid-1800s,

Settlers cooked sap in large cauldrons hanging on tripods over open fires.

flat-bottomed pans and firebox arches had become popular. Originally developed to process sorghum syrup, the Cook Evaporator was patented in 1858. Improvements were, and continue to be, made to original evaporator designs. Even with modern technology, however, the fact remains that sap must be boiled and evaporated by heat to create the distinct color and flavor of pure maple syrup.

---

## *Commitment and Return*

The syrup-making season is a short but busy one. If you have 50 taps and want to produce the most syrup possible, you'll need to commit at least 10 hours of work per day when the sap is running. Sap doesn't run *every* day during the syrup-making season, but in an average season you can collect sap for 10 to 15 days over a 4- to 6-week period. If a 10-hour syrup-making day seems impractical for you, lessen your workload. You need to make only as much syrup as you want. It's not uncommon for people with as few as one or two yard maples to make a small amount of maple syrup. If you've collected more sap than you want, you may be able to sell it to larger syrup producers or give it away. Here's how much syrup you can expect to produce in an average season:

- **Per tap,** 1 quart
- **From 50 taps,** 12.5 gallons
- **From 150 taps,** 37.5 gallons
- **From 500 taps,** 125 gallons

---

CHAPTER 2

# IDENTIFYING AND TAPPING MAPLE TREES

Maple trees grow naturally in the forests of eastern North America, from the coastline of southeastern Canada and the northeastern United States westward through Ontario and Minnesota. To the south they can be found in Georgia, continuing as far west as eastern Kansas and Oklahoma.

There are 13 native maple species in North America, all of which can be tapped: sugar, black, red, silver, box elder, mountain, striped, bigleaf, chalk, canyon, Rocky Mountain, vine, and Florida. Only four of those are commonly tapped: the sugar, black, red, and silver maples; their sap has a higher sugar content than sap from thc other species.

Maple forests are located in the northeastern United States and southeastern Canada.

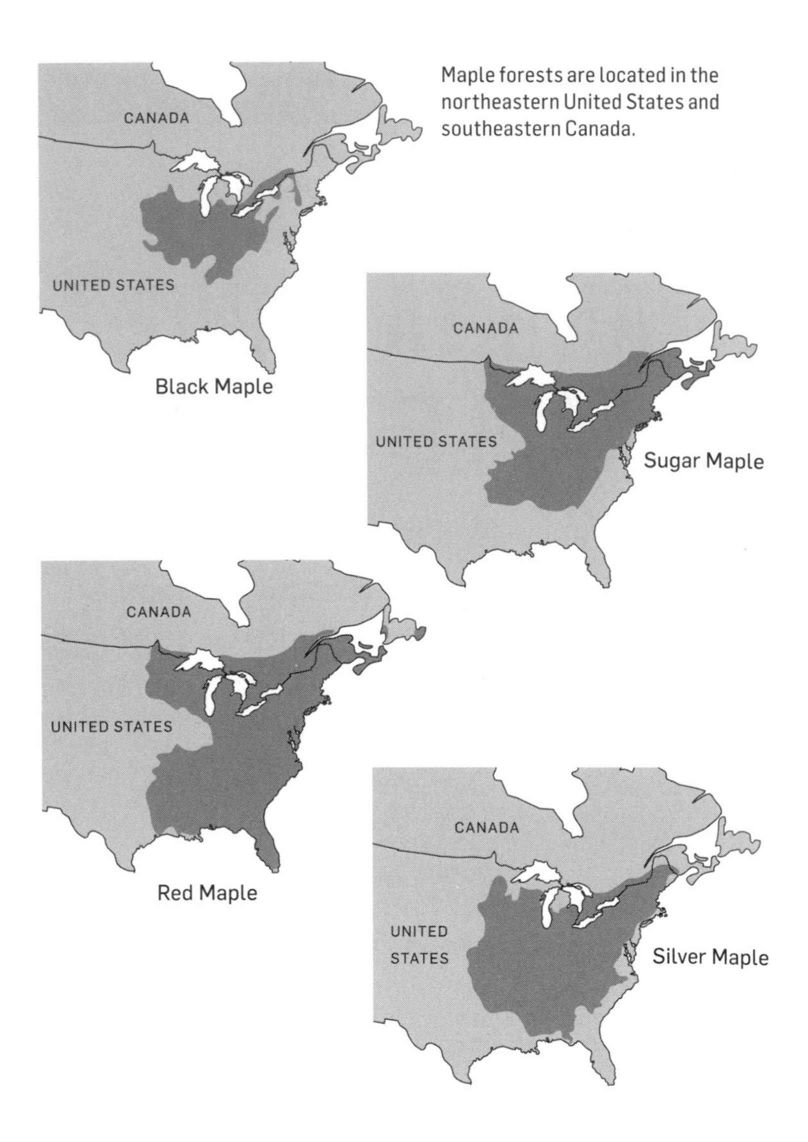

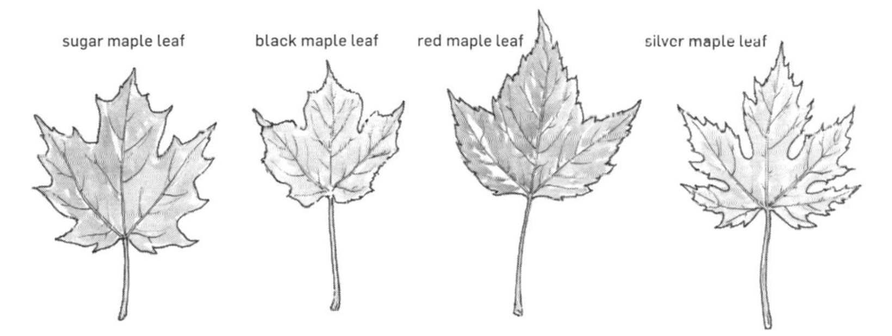


Four types of maple trees are commonly tapped: sugar maple, black maple, red maple, and silver maple.

## *Let Trees Grow before Tapping*

It is important to tap more mature maple trees and leave the smaller ones to develop and grow into healthy sap producers. A wound in a larger tree will do less damage than a wound in a smaller tree because a larger tree has more surface area, and the damage is somewhat displaced by the size of the tree. Smaller trees also have less sap.

There are no stringent guidelines to follow, but a good rule of thumb is to tap trees that are larger than 10 inches in diameter at chest height. Tapholes should not be plugged after the spout is pulled out at the end of the sap-collecting season. Just as your body heals a cut, the tree will naturally heal, and the taphole will eventually close.

## HOW TO IDENTIFY A MAPLE

THE EASIEST WAY TO IDENTIFY A MAPLE TREE is by its leaves; it is difficult to identify a maple without its leaves, but it can be done. Mature maples are large dense trees with rounded crowns. The trunks are usually straight and free of branches for two-thirds or more of the height. The bark of a mature maple is usually light to dark gray with vertical smooth strips that may curl. Look for a tree whose branches, buds, and leaves are located opposite one another, and on which the twigs are arranged in pairs on opposite, rather than alternating, sides of a branch. During the growing season the buds and later the leaves are arranged in pairs on either side of the twigs.

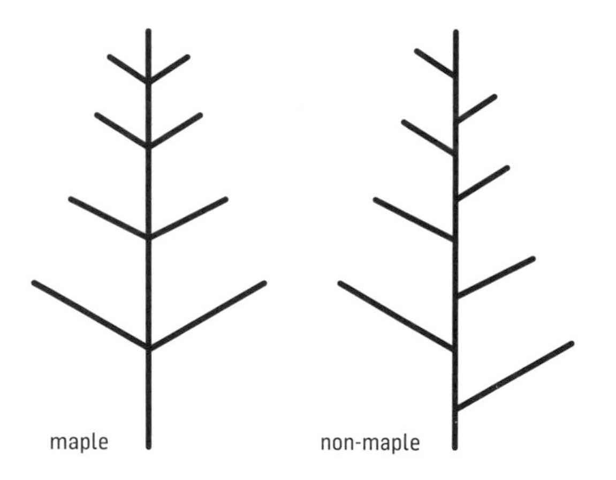


When identifying maple trees, look for branches that are in opposite (left) position rather than alternate (right) position.

Because it can be hard to identify maple trees without leaves, it is most efficient to identify and mark your trees in the summer or fall, when you can easily determine what kinds of trees you have by looking at the leaves. For marking you can use either tree-flagging tape or a forestry-approved marking paint. Flagging tape, found at hardware stores, is a nonadhesive tape, usually brightly colored, that can be tied around a tree to mark it. Forestry-approved marking paints are designed to withstand harsh weather but are not harmful to trees (see Resources for suppliers).

A mature maple tree has a rounded crown.

## HOW TO MAINTAIN HEALTHY TREES

LANDOWNERS CAN UTILIZE FOREST MANAGEMENT strategies to maintain a healthy woods; however, it is most important to allow a woods to maintain its natural diversity. For example, if your woods has a variety of tree species, you don't want to cut all but the maples. Leave the forest as it developed naturally, though you may want to thin your woods and clear dead trees, as this allows for younger ones to grow.

When thinking about sugar bush management, keep your objective in mind. What do you want your sugar bush to accomplish and what can you do to achieve this goal? It is best to ask your local forester for advice. If you don't want to contact a forester, at least talk to an experienced syrup producer who has an operation similar to the type you hope to establish.

There are also good forestry maintenance resources available online. Search for "forestry maintenance" in your state to find resources specific to the area in which you live. The resources don't have to be specific to a sugar bush; they just have to focus on hardwood forests in general.

## WHEN TO TAP

THE TIME TO TAP MAPLE TREES IS in late winter or early spring, when the weather is above freezing during the day and below freezing at night. Sap generally runs from early February to late March in the more southern states and from early March to mid-April in the northern states and provinces. The exact calendar date varies greatly from region to region and year to year.

Begin by watching the forecast, and look for an extended period of desirable weather. One day of perfect weather may cause excitement but may not lead to an extended freeze-thaw cycle that will cause the sap to run. In an average year you will have 10 to 15 days when the sap runs.

## Not Too Early and Not Too Late

Tapholes begin the process of healing as soon as they are drilled, so if you tap your trees too early, you risk missing out on sap flow later in the season. The tapholes may start drying out before the sap is done running.

### *Bacterial Growth and Maple Grades*

The same bacteria that cause a taphole to heal will cause you to make darker syrup as the season progresses. At the beginning of the syrup season, when the taphole is fresh and your equipment is clean, you will probably make syrup that is lighter in color, often referred to as Grade A Light Amber. As the season progresses, and bacteria collect in the taphole and on equipment, you will probably make syrup that is darker, often referred to as Grade A Medium or Dark Amber. Really dark syrup, sometimes with an off, or atypical, maple flavor, is considered Commercial grade syrup. This bacteria growth is a natural occurrence and is nothing to be concerned about.

By tapping your trees too late, on the other hand, you may miss the early sap flows. Syrup produced earlier in the season is usually higher quality (with a lighter color and a more delicate maple flavor) than syrup that is produced later in the season. This is due to warmer temperatures, natural bacterial growth, and equipment that may not be as clean as it was at the beginning of the syrup season. Later in the spring, trees also begin to go through a chemical reaction as they start to bud. Sap that is collected after that point can also make lower-quality syrup.

## Helpful Tips

Watch what the experts in your area are doing. People who have been tapping trees for a long time know what to look for in local weather patterns. They watch for a forecast with daytime temperatures reaching into the 40s (above 4°C) and nighttime temperatures plunging below freezing. Don't rely on national weather outlets. Your forecast must come from local sources because current, local weather greatly impacts the sap run.

Also, temperatures taken in wooded areas are colder than those taken out in the open. The difference can be as much as 8°F (4°C). If you're considering tapping on a 40°F (4°C) day, the temperature in the woods may still be 32°F (0°C). Tap one tree that is representative of your lot, as a "test tree." When the sap starts running in that tree, you can tap the rest.

---

# WHAT YOU NEED

You will need a drill and bit, a spout, and a hammer. A spout (also called a tap, spile, or spigot) is a strawlike fitting that is inserted into the tree. The spout will allow sap to drip out of the tree and into whatever container you use for collection.

## Hooked vs. Hookless Spouts

Spouts come in a variety of styles and are typically made of aluminum alloy, stainless steel, or a plastic composite. Spouts come with or without hooks. Hooked spouts are generally used with the flat metal lids that cover sap pails. The hooked spout attaches to the lid via a thin metal rod that feeds through a hole

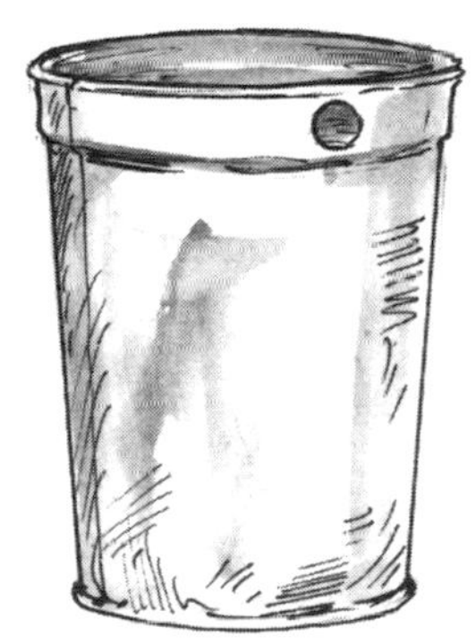

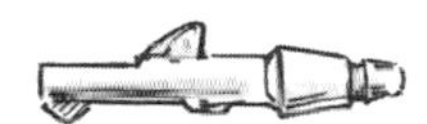

The basic tools for tapping a tree include a metal sap bucket, a spout, and a hand-brace drill and bit.

A hooked spout works with a flat sap pail lid. The lid and the spout are fitted together, and the sap bucket hangs from the hook on the bottom of the spout.

on the top of the spout (see illustration on page 15). When the spout is in place on the tree, a pail hangs from the hook on the underside of the spout. If there is not a hole through the top of a hooked spout, you need to use a roof-type metal pail cover that mounts to the rim of the syrup pail.

Hookless spouts are a bit more versatile than hooked spouts, but they do not work with flat pail lids. Hookless spouts have a notch on the top over which you hang your pail or Sap Sak holder. Hookless spouts work very well with metal pails and roof-type metal pail covers, Sap Sak holders, and plastic 5-gallon pails and their plastic covers — as well as with other more makeshift collection vessels.

## Spout Sizes

The maple industry has done a good job of standardizing spout sizes. The two most common sizes are 7/16 inch and 5/16 inch.

Historically, the 7/16-inch spout was most commonly used, but research suggests that a taphole will heal faster if a 5/16-inch spout is used. Research has also shown that spouts smaller than 5/16 inch restrict sap flow and lower production yield. A 5/16-inch spout has almost exactly the same sap output as a 7/16-inch spout.

Use the same size drill bit as your spout. For example, if you are using a 7/16-inch spout, use a 7/16-inch drill bit. For a 7/16-inch spout, drill a hole, at a slightly upward angle, that is 2 to 2½ inches deep. For a 5/16-inch spout, drill a hole that is 1¼ to 2 inches deep. Make sure to use a sharp wooden drill bit, and clear the hole of any debris before putting your spout in place.

Both sizes of taps are commonly sold, and which is best for you is a matter of personal preference. Often a customer will select a particular size based on the size of the drill bit she already owns. If a customer is starting from scratch, he will often choose the 5⁄16 size strictly because he likes the idea that the tree heals faster with the use of the smaller spout. Either size is absolutely acceptable. Neither spout does any lasting damage to the tree.

## Collection Vessels

Whatever you use to collect sap can hang directly from your spout. You don't need to nail it to the tree. Metal buckets with covers, 5-gallon pails, or Sap Sak holders are most commonly used (all discussed in chapter 3), but other items will work as well. See chapter 7 for information on tubing for larger operations.

**Shopping List for 50 Taps**

- 50 Sap Sak holders and bags, or 50 pails with covers
- 50 spouts, either hookless or hooked, depending on application
- Drill and bit to match spout size
- 5-gallon pails with handles for gathering and hauling sap, one for each person collecting sap
- Gathering tank to carry sap from woods; could be mounted on a tractor or all-terrain vehicle

*List continues on following page*

- Transfer pump to transfer sap from gathering tank to holding tank
- Holding tank to feed your cooking pan; any hoses and connections needed
- 2-foot-by-3-foot pan for cooking sap (for more information on sizing your pan, see chapter 9). We suggest a flat-pan evaporator with a flat pan, an arch, and a smokestack. To save money you could purchase just the pan and build your own firebox and stack.
- Syrup hydrometer
- Hydrometer cup
- Thermometer for finishing and bottling syrup
- Smaller finishing pan that could double as a bottling pan
- Propane-powered burner
- Orlon bag or sheet and paper prefilters
- Valve for a finishing or a bottling pan
- Glass or new plastic bottles with new covers
- Labels

## PICKING YOUR SPOT

As mentioned, your trees should be at least 10 inches in diameter at chest height for you to tap them.

Choose a smooth spot on the tree that doesn't have any visible scarring and is at least 6 inches away, horizontally, from any old visible tapholes. Tree scarring, caused by punctures to a tree (such as tapholes), impedes the flow of sap and runs mostly vertically through trees. You never want to tap directly above or below an old visible taphole. Don't worry about tapping too close to old tapholes that have healed completely. Younger, healthy trees will heal faster than older trees. Tap the tree at a comfortable working height.

On a large, old tree, you may need to remove some loose bark to create a smooth spot to tap. You may put more than one tap on large trees, but taps should be kept 18 to 20 inches apart horizontally.

You may tap any side of the tree; it is a myth that you need to tap only the south-facing side. To avoid old tapholes from year to year, you will have to make your way around the tree. Some experts suggest tapping in a systematic spiral pattern around the tree.

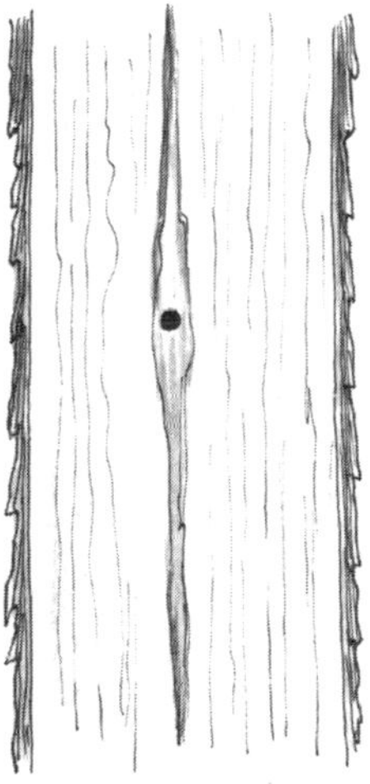

Tree scarring runs mostly vertically within maple trees. Do not tap above or below old tapholes to avoid tree scarring, which can impede sap flow.

## DRILLING AND TAPPING

USING A POWER DRILL OR HAND BIT and brace, drill a hole that is about 2 inches deep at a slightly upward angle. If using a power drill, be careful not to let the bit spin and cauterize (heat and begin to seal) the hole. Cauterizing a hole will limit the sap output.

Place your spout into the hole, and gently tap it in about 1 inch with a hammer until it is snug. There should be about a 1-inch gap between the back of the drilled hole and the end of the spout once the spout is tapped into place. This open space in the taphole is sometimes referred to as the "collection area." The collection area allows for a small amount of exposed wood grain where the sap can come to the surface. The sap also gathers in the collection area and creates a small amount of pressure, which pushes the sap from the tree.

Be careful not to drive in the spout too far. This could crack the wood, and the crack could leak sap. If you've cracked the wood, you may see sap running down the side of the tree, and your spout may be dry. There is nothing you can do to fix this, but if it does happen, learn from the mistake. The tree will heal itself.

Hold your drill steady and drill about 2 inches into the tree. Make sure the hole is clear of debris before pounding in your spout.

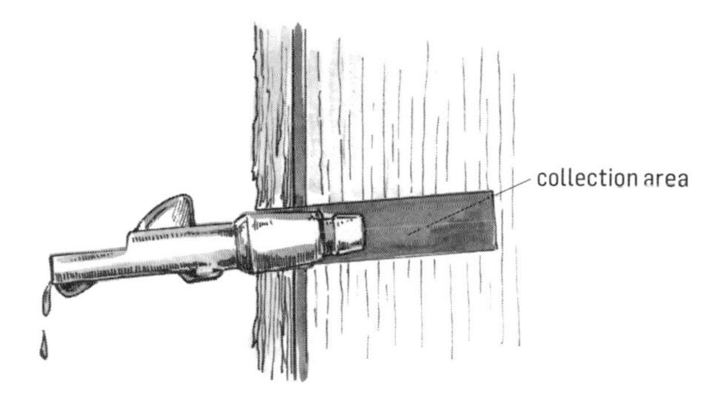


Leaving a 1-inch open space inside the taphole, between the back of the drilled hole and the end of the spout, allows for the sap to gather in the hole and create pressure for even sap flow.

## COLLECTING SAP

AFTER THE SPOUT IS IN PLACE, hang a bucket or Sap Sak holder on it and wait for the sap to start dripping. Be patient. It may not flow right away, but when the weather is just right, you'll be able to watch the sap drip from the spout.

You will probably average 10 to 15 days of actual sap production in a season. The amount of sap a tree will produce on any given day depends greatly on the freeze-thaw cycle, the barometric pressure, and the amount of thawed ground moisture in your woods. The barometric pressure falls when a storm approaches, and a large change in barometric pressure can lead to a good sap flow. That's why the sap runs faster when it snows or rains in the spring.

Hang your bucket directly from the spout in the tree.

On average one taphole will yield about 1 gallon of sap a day. It takes about 40 gallons of maple sap to make 1 gallon of maple syrup.

CHAPTER 3

# GATHERING SAP

What will you hang on your trees to collect sap? Metal sap buckets, Sap Sak holders, and 5-gallon plastic pails are all good options. Whatever you choose to use, it should be a container that is acceptable for storing potable water. Potable water storage containers can be found at most hardware and home improvement stores.

---

## COLLECTION VESSELS

THE MOST DESIRABLE VESSEL for collecting is a 16-quart metal bucket with a roof-type or flat cover designed for gathering sap. A container this size can be lifted from the spout and easily dumped into a larger gathering tank. Purchasing this type of bucket new has become rather expensive. Good alternatives are Sap Sak holders and 5-gallon plastic pails.

## Metal Buckets

Metal buckets have been the traditional method for collecting maple sap during the modern era of syrup making, and they are still one of the best containers for the task. In recent years, however, the high cost of new metal buckets has made them one of the more expensive vessels for sap collection. If you can find used metal buckets for sale at a reasonable price, they are worth purchasing.

Pail covers come in either roof-type or flat styles.

## Sap Saks

A Sap Sak holder is a very economical option for gathering sap. As a bonus it can easily be stored in a small space, whereas sap pails need to be stacked for storage and, depending on how many you have, can take up a lot of space. The sak holder hangs on either a 5⁄16-inch or 7⁄16-inch hookless metal spout. The unit consists of three parts: the bag (usually sold separately), the holder, and the inner ring. The bag holds about 4 gallons of sap. It will need to be replaced every year. The holder can be reused from year to year indefinitely.

You will typically need one bag per holder during your period of sap collection. It is a good idea to have some extra bags just in case one rips or is punctured. Squirrels like the sweet taste of the sap and sometimes try to bite through Sap Saks. If squirrels become problematic, try hanging duct tape off the corners of the sap bags. Squirrels do not like the taste of the duct tape, and this will prevent them from trying to bite the corners of the bags.

## ASSEMBLING A SAP SAK HOLDER

Sap saks and holders are a great alternative to traditional metal sap buckets. They work great; however, some assembly is required. Once you get the hang of it, it doesn't take long to assemble them, and you can do it indoors, where it is warm. You may want to wear work gloves while putting them together, as the edges of the holders are sometimes a little sharp.

1. Remove the inner ring from the holder. Place the top of the bag inside the ring so it extends about 4 inches above the ring.

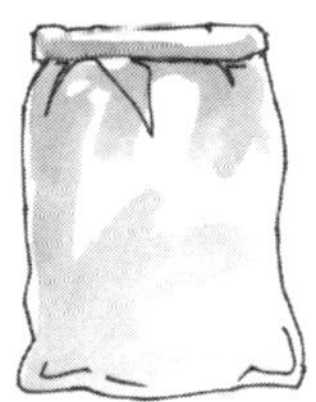

2. Fold the bag completely over the ring, and tuck about 1 inch of the bag back up under the ring.

*Steps continue on following page*

3. Place the ring back into the holder, and pull down firmly on the bag to make sure that the ring has been seated in the V-notches of the holder.

4. Now the Sap Sak holder is ready for the tree. Slide the hole in the holder over the spout, and wait for the sap to flow.

## Plastic Pails

If you use 5-gallon plastic pails, make sure they are acceptable for storing potable water. New potable water-grade 5-gallon plastic pails can be purchased from hardware or home improvement stores. Drill a 1-inch hole in the side of the pail about 2 inches below the rim. Hang the pail from a hookless spout the same way you would hang a metal bucket or Sap Sak holder. Put the cover on the pail. The pail can then be lifted off the spout, and the sap can be poured out of the hole you drilled in the side of the pail.

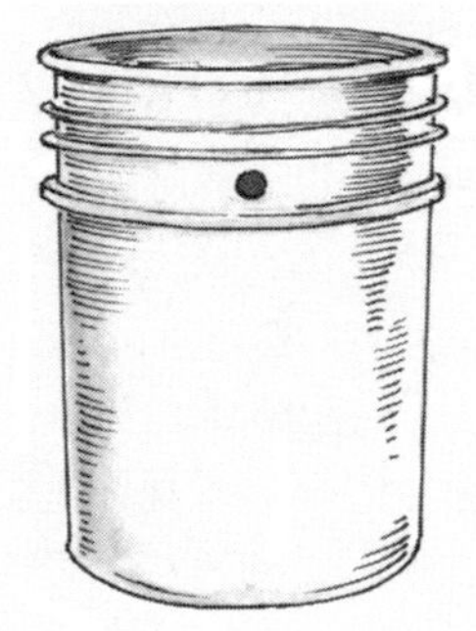

Collecting sap in 5-gallon plastic pails is a good alternative to purchasing new metal buckets. Drill a hole in the side of the pail, and hang it directly on the spout.

## AVOID TUBING WITH BUCKETS

It has become very popular to collect sap using inexpensive plastic tubing spouts, with a short length of tubing running to a bucket sitting on the ground. This is not something we recommend. Tubing and tubing fittings are meant to be used with a gravity or vacuum tubing system.

There are a variety of problems with collecting sap via tubing into a bucket:

1. **Plastic is porous and holds bacteria.** Plastic tubing and fittings are nearly impossible to clean. Plastic spouts are meant to be disposable, and most large producers who collect sap with a tubing system replace their plastic tubing spouts each year. Plastic should not be cleaned with soap or bleach. Even when rinsed very well, it will retain residue from soap or bleach. Producers using tubing systems flush their lines at the end of the year with water. At the beginning of the next syrup season, to flush mold and debris from the lines, they will dump the first several hundred gallons of sap before they actually begin collecting it. Hobbyists collecting sap from just a few trees cannot afford to dump gallons of sap.

   If you're going to collect sap this way, you should replace your plastic spouts and tubing each year. After only 2 or 3 years of doing this, you will negate your initial cost savings. Over the long run it is less expensive to purchase metal spouts.

2. **The sap flow is restricted.** Even when properly used, tubing causes a restriction of sap flow. Therefore, you may collect slightly less sap than if you used a metal spout that drips directly into a bucket. When sap is gathered with a properly installed tubing system, the efficiency of the system makes up for any sap flow restrictions.

3. **Buckets sitting on the ground may tip over.** Perhaps not all your buckets will fall over, but snow melts; animals, people, and machines move through the woods; and the wind blows, sometimes hard. Any of these scenarios could knock over your buckets. Also, if your length of tubing is too short, it is likely to blow out of the bucket. If you are collecting at the end of the day and find an overturned pail or tubing hanging outside a bucket, you've missed an entire day of sap run.

4. **Tubing can act like a straw.** Say there has been a good sap run and your bucket is full. Your length of tubing is now sitting below the sap line in your pail. If your tube is 28 inches or shorter, it can act like a straw and begin to draw your sap back up into the tree once the tree starts to freeze. Your tubing has now become counterproductive. If you are going to collect sap this way, make sure your tubing is at least 30 inches long to avoid this drawback.

5. **Tubing and pails don't mix.** Additionally, plastic tubing spouts are not meant to be used with pails. Customers

are attracted to the idea of using plastic tubing spouts to hang their buckets because they are much less expensive than metal bucket spouts, but plastic tubing spouts may not hold the weight of a full sap bucket. Just as plastic tubing spouts should not be used with sap buckets, metal bucket spouts are not meant to be used with tubing.

## TESTING SAP

YOU DO NOT NEED TO TEST THE SUGAR CONTENT of your sap if you are producing syrup. However, if you are going to sell your sap to another syrup maker, or if you are just curious, you will need to do so. The easiest way to test sugar content is with a sap hydrometer (different from a syrup hydrometer).

Producers buy sap by the gallon and pay a certain rate according to sugar content. Large syrup producers may buy sap because they have a big evaporator that can handle large amounts of sap, and they want to spend more time cooking sap and less time collecting it. Someone who just enjoys collecting sap can make some money by selling it and doesn't have to invest in syrup-making equipment.

Sometimes a syrup producer will work on a cooperative model, trading finished syrup for raw sap. Sometimes they will pay outright for the sap. Either way, both parties will want to know the sugar content of the sap for the purpose of assigning it a proper monetary value.

## Using a Sap Hydrometer

Sap hydrometers are calibrated to work with cold sap (about 38°F [3°C], or average outside temperature during syrup season). If you are testing sap on a very warm day, or if your sap sample came from a bucket hanging in direct sunlight, you may not get an accurate reading. Just be aware of this when testing the sugar content of your sap.

1. Set the sap hydrometer in a full cup that is almost as tall as the hydrometer, or set it in a pail or tank of sap. Hold the hydrometer tightly until it is floating on its own. A hydrometer released in sap can hit the bottom of the cup or pail and crack the glass. This is a *very* fragile instrument. If the hydrometer does break while you are testing your sap, discard the sap with the broken glass in it.
2. When the sap hydrometer is floating, read the number that is floating at the surface of the sample. This is your sugar content. Average ranges are from 2 to 3.5 percent.

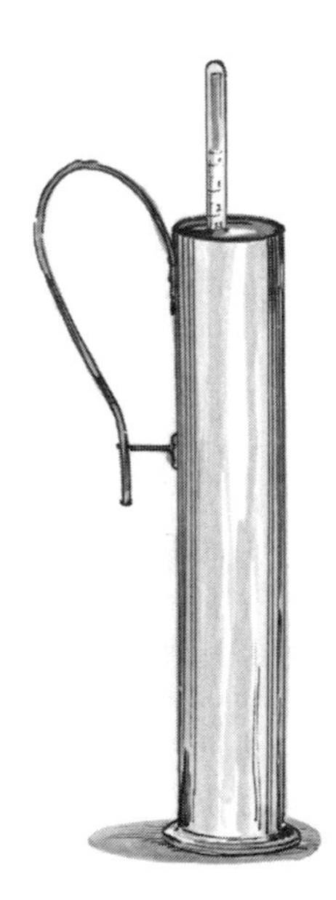

This sap hydrometer shows that this sap has a sugar content of about 2.5 percent.

### Rule of 86

It takes approximately 40 gallons of sap to make 1 gallon of pure maple syrup. By using the Rule of 86 (86 divided by the sugar content of your syrup), you can determine more closely how many gallons of your sap it will take to make pure maple syrup.

For example, if the sugar content of your sap (taken from the sap hydrometer) is 2.5 percent, it will take about 34.4 gallons of your sap to make 1 gallon of pure maple syrup:
$86 \div 2.5 = 34.4$ gallons.

---

## FILTERING SAP

YOU USUALLY DO NOT NEED TO FILTER YOUR SAP. Unlike syrup, which has sugar sand (the grainy sediment that is created every time sap or syrup is boiled) that needs to be removed before it is bottled, there is nothing in sap that needs to be removed before you cook it. The covers on the pails should do a pretty good job of keeping bark, insects, and small animals from falling into your sap. However, if you find that your covers aren't cutting it and there is more debris in your sap than you would like to cook, you can strain your sap through a light fabric or screen to remove the undesirable items.

Maple syrup supply dealers sell a variety of items for filtering sap. Many of these items work with more industrial-size pumps for filtering large quantities of sap in short periods of time and are geared toward large syrup producers. For small producers we often recommend using a paper prefilter. Prefilters are sold with Orlon for the purpose of filtering syrup (discussed in chapter 5), but they work quite well for filtering sap, and they are very inexpensive. Cheesecloth, window screen, and cotton T-shirt material also work for filtering sap. If you are filtering with a recycled item, make sure it is clean. Boil it to remove any soap residue.

---

## FROM TREE TO PAN

Once you have collected some sap, you'll need to get the sap from the trees to your cooking location. You can either carry the sap out in the buckets in which you've collected it or empty the buckets into a larger container, then haul that out of the woods. Any containers used to store sap should be potable water–grade. New plastic potable water–grade holding tanks can be purchased at most farm supply stores or from a maple syrup supply dealer.

If you are using a barrel or some other collection tank, you may want to load it onto an all-terrain vehicle or some other kind of vehicle that can navigate easily through the woods. You'll need a transfer pump to move the sap from your collection tank to an elevated tank that will feed to your cooking pan or evaporator.

Check your buckets or Sap Saks every day, and gather your sap as needed. Some days your buckets may be full, and other days there may be very little sap or no sap at all. One advantage of using Sap Sak holders is that you can see from a distance if there is sap in the bag. With pails you have to look directly into them to see if they are holding any sap.

CHAPTER 4

# COOKING SAP

Now that you have some sap, all you need to do is boil it to evaporate the excess water. Thinner stainless steel works best for heat transference, and any heat source that will boil water will boil sap.

Cook the sap promptly after you've gathered it. As spring temperatures rise, sap can spoil quickly; sap that has spoiled will have a foul odor and look milky. If you add spoiled sap to a pan of sap that is cooking, you can ruin an entire batch of syrup. Syrup that has been made from spoiled sap will be stringy and off-flavored.

If you have to store your sap because you cannot cook it right away, keep it as cold as possible, preferably in a refrigerator or freezer. You can throw out any ice that gathers on top of frozen sap, as it is mostly water.

## BOILING: PANS AND BURNERS

Syrup makers often use a flat pan and a wood fire for cooking sap. To beginners who have 1 to 10 taps and don't want to invest in a stainless steel syrup pan, we often suggest a home turkey fryer, which typically comes with a 30-quart cooking kettle and a propane burner. Turkey fryer cooking kettles don't have as much surface cooking area as a flat syrup pan, but they can be purchased at most home improvement stores and are reasonably priced. Whatever pan or container you use, make sure that it is food grade.

If you can keep a close watch over the cooking process, fill your pan with about 2 inches of sap. This will allow for an optimum boil. Add your heat source, and continue adding small amounts of sap as the water evaporates to maintain this 2-inch level.

If you can't closely monitor the cooking, fill the pan with enough syrup that it won't go dry while you are away. Come back and check it often. Never let any part of your pan that comes in contact with direct heat go dry — your pan and syrup will scorch and burn.

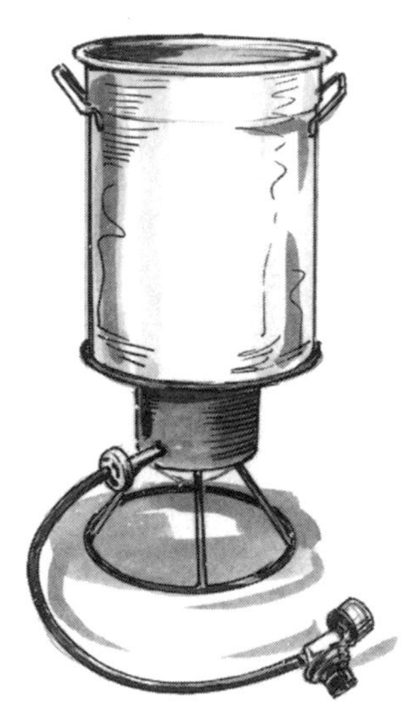

Cooking with a turkey fryer is a convenient way to try making maple syrup without investing a lot of money in a syrup pan or small evaporator.

You can't apply too much heat to the pan as long as there is liquid in it, but remember that the longer your sap is over the heat, the darker your finished syrup will be. Darker syrup doesn't mean lower quality; it just means that the maple flavor might be stronger.

## FINISHING YOUR SYRUP

When your sap has reached a tenth of its original volume or has nearly reached 66 Brix (see Using a Syrup Hydrometer, page 36), you can stop adding sap. You will need to finish your sap at this point. Some syrup makers prefer to finish their syrup in a smaller pan on a heat source that is easier to control than an open fire, such as a camp stove or propane burner.

Whichever your heat source, watch your sap very closely. Some foam may develop in your pan; if it does, a drop of defoamer (see box below) will prevent the foam from getting too high. Also, make sure that the liquid in your pan is still

### Defoamer

*Defoamer is typically a blend of various vegetable oils formulated specifically for the purpose of making syrup. The oil breaks up the surface tension of the boiling sap and allows the heat to escape faster, thus breaking down the foam. Defoamer can be purchased from a maple syrup supply dealer.*

deep enough that, as you finish cooking, your pan will not develop a dry spot and burn.

You should test your syrup at this point with a thermometer, a syrup hydrometer, or a refractometer (see Using a Refractometer, page 42).

---

## USING A SYRUP HYDROMETER

SAP TURNS INTO SYRUP when it is 7°F (4°C) above the boiling point of water. The temperature at which water boils is affected by elevation and barometric pressure. Water typically boils at 212°F (100°C) at sea level, and sap typically turns into syrup at 219°F (104°C), but you will need to test these temperatures for yourself. Each time you are about to finish a batch of syrup with a thermometer, boil some water, note the temperature at which it is boiling, and add 7°F (4°C). When your sap reaches this temperature, you have syrup.

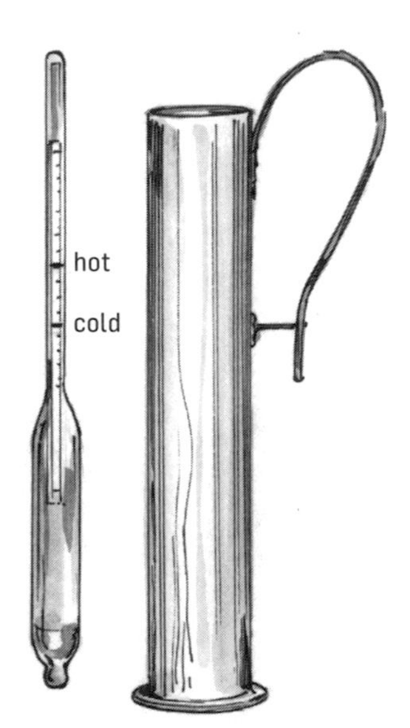

Using a syrup hydrometer is the easiest and most accurate way to determine if your sap has turned into syrup.

Using a syrup hydrometer eliminates the need to boil water and figure out the temperature for finishing your syrup. With a hydrometer you read either the red "hot test" line or the red "cold test" line, depending on the temperature of your syrup.

A syrup hydrometer (different from a sap hydrometer) uses either the Brix or the Baume scale, both of which are used to determine the density of maple syrup. The Brix scale (more commonly used today) compares the density of maple syrup to that of a sugar solution with a known percentage of sugar. The Baume scale relates the density of syrup to a salt concentration of the same density. When the liquid measures 66 Brix or 32 Baume, it is syrup and should be removed from the heat.

Be very careful when handling a hydrometer; it is a fragile glass instrument. You may want to keep a spare on hand in case of breakage.

## Incremental Testing

When you are finishing your sap, take occasional temperature readings using a thermometer to see when you are close to having syrup. When the temperature of your boiling sap reaches 5°F (3°C) above the boiling point of water, start checking it every 10 minutes with a hydrometer. The closer you get to 66 Brix, or 32 Baume, the more often you need to check.

The chart on the following page shows how incremental temperatures above the boiling point of water relate to Brix measurements on your way to making syrup. For example, if your thermometer reads 3°F (1.7°C) higher than the temperature at which your water boils, your sap is at 49.0 Brix. That's well below the 66 Brix required for pure maple syrup.

### Temperature and Brix Equivalents

| °F (°C) ABOVE BOILING POINT OF WATER | BRIX |
|---|---|
| 3.0 (1.7) | 49.0 |
| 4.0 (2.2) | 54.9 |
| 5.0 (2.8) | 59.7 |
| 6.1 (3.4) | 63.4 |
| 6.9 (3.8) | 65.5 |
| 7.1 (3.9) | 66.0* |
| 7.3 (4.1) | 66.5 |
| 8.0 (4.4) | 68.0 |
| 9.1 (5.1) | 70.0 |

*At 66 Brix, you have achieved maple syrup.

Note: A syrup hydrometer is calibrated to read cold finished syrup at 60°F, or hot finished syrup at 211° (16°C). Don't try to find a temperature reading for 219°F (104°C) on the hydrometer because it's not there. It is calibrated to take into account any cooling that may happen while you fill your hydrometer cup. For hot syrup, you'll be drawing off when the red "hot test" line on the hydrometer floats evenly with the surface of the syrup.

## *Troubleshooting to Avoid False Readings*

If the syrup temperature is significantly different from the hydrometer's calibrated temperature, it could be due to one of the following reasons:

- **There are sugar sand or calcium deposits on the hydrometer.** Sugar sand deposits are the grainy sediment that is created every time sap or syrup is boiled. Calcium deposits look similar to hard water deposits that sometimes form on a faucet. Carefully remove the deposits by scraping them with a knife blade or by soaking the hydrometer in a lime-removing cleaner. Rinse very well.
- **There is syrup sticking to the stem above the red line.** This, and sugar sand deposits, both add weight to the instrument and vary its reading. Remove syrup from the hydrometer by rinsing it well with hot water. Wipe the hydrometer dry with a clean rag to ensure you've removed all the syrup residue.
- **The paper scale inside the hydrometer stem has loosened or shifted position.** This is rare but does occur from time to time. Always take note of the paper's location so you know if this happens. Try marking it with a permanent marker, and take note of the marking each time you use the hydrometer.

## Hot Test

For the hot test take a sample of syrup directly from the boiling pan. Do not let the syrup sit or cool. Sap typically turns to syrup at 219°F (104°C), but the syrup will cool several degrees in the hydrometer cup, so the hydrometer is calibrated for 211°F (99°C).

1. Fill a hydrometer cup at least 8 inches deep with hot syrup.
2. While holding the cup over the boiling pan, slowly lower the hydrometer into the cup. The syrup may overflow when you lower the hydrometer into it. Be sure to hold onto the hydrometer until it is floating on its own or resting on the bottom of the sample cup — if the syrup is still thin, the hydrometer can hit the bottom of the metal cup when released and may crack. If the hydrometer breaks while you are testing your syrup, discard the syrup with broken glass in it.

Here's how to read the hydrometer:

- **Hot Test red line is even with the syrup's surface.** Your syrup is finished. Remove it from the heat.
- **Hot Test red line floats above the syrup's surface.** Your sample is too heavy. Add a small amount of hot sap to your syrup to thin it and retest. If your syrup is still too thick, continue adding small amounts of hot sap and retesting until your hydrometer reads 66 Brix or 32 Baume.

- **Hot Test red line is buried in your syrup.** The syrup is too thin, and you need to continue to boil it.

## Cold Test

For the cold test you'll need to use syrup that is room temperature. The Cold Test line on the hydrometer is calibrated to 60°F (16°C).

The general procedure for the cold test is the same as for the hot test:

1. Fill a hydrometer cup at least 8 inches deep with room-temperature syrup.
2. Hold the cup over the syrup pan while you gently lower the hydrometer into the cup. Hold the hydrometer until it is floating on its own or resting on the bottom of the cup. If the syrup is too thin, the hydrometer can crack when it hits the bottom of the cup.
3. Allow some time for the hydrometer to stabilize in the cold syrup. If the syrup is finished, the Cold Test red line on the hydrometer will be even with the surface of the syrup.

If your syrup is too thin, bring it back to a boil, and continue cooking and retesting until it reaches 66 Brix or 32 Baume.

If your syrup is too thick, bring it back to a boil. Add a small amount of sap to your syrup to thin it, and retest. If your syrup is still too thick, continue adding small amounts of sap and retesting until your hydrometer reads 66 Brix or 32 Baume.

## USING A REFRACTOMETER

A REFRACTOMETER IS ANOTHER TOOL used to measure the sugar content in sap and syrup. A refractometer measures the refractive index of the syrup, or how light bends as it passes through the syrup. There are some advantages to using a refractometer rather than a syrup hydrometer, but a refractometer is considerably more expensive. To use one:

1. Slightly cool the syrup you are going to test. If you use boiling syrup, the small amount of hot syrup will cool quickly and sometimes blur the reading. To cool the syrup, transfer 1 or 2 ounces of your boiling syrup into a small glass container, such as a 2-ounce sample jar. Screw on the cap so moisture doesn't escape while it cools. This small amount will cool quickly (about 10 minutes). The syrup should be warm to the touch but not so hot that it will burn your skin.
2. Shake the sample, in case any additional evaporation caused condensation to form on the inside of the cap, and draw some of the syrup into a pipette.
3. Put a drop of the cooled liquid on the small window on one end of the refractometer.
4. Look through the eyepiece. The drop of sap appears as a shadow against the calibrated, numbered background scale. Most syrup refractometers measure in Brix. Finished syrup should measure at 66 Brix.

When measuring hot syrup, the only real benefit to using a refractometer rather than a hydrometer is that less syrup is required for the refractometer. The instruments are equally accurate.

When measuring room-temperature syrup, a refractometer works more quickly than a hydrometer. It takes time for a hydrometer to stabilize in cold or room-temperature syrup. With a refractometer, there is no wait for an accurate measurement.

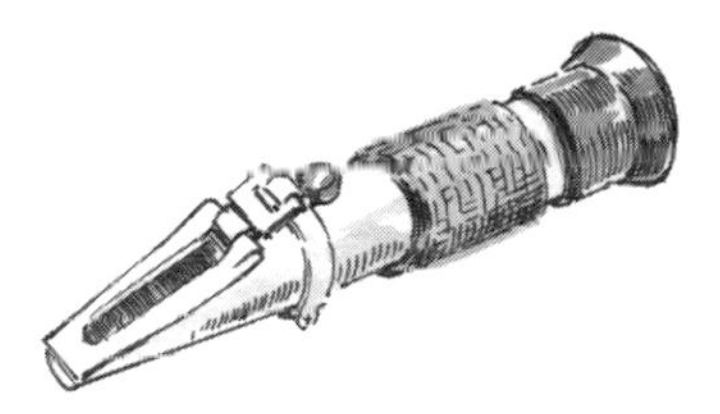

A refractometer is another tool for testing the sugar content of sap and syrup.

## Is It Okay to Use an Alcohol Hydrometer?

*We are often asked by customers if they can test their syrup with the same hydrometer or refractometer they already own for making alcohols such as wine or beer. The instruments for alcohol making and syrup making are calibrated differently. It may be possible to convert the numbers, but it's probably quicker and easier to have a hydrometer specific to the task. Syrup hydrometers and refractometers can be purchased from a maple syrup supply dealer.*

CHAPTER 5

# FILTERING AND BOTTLING SYRUP

Finished maple syrup should be filtered to remove sugar sand — the grainy sediment that is created every time sap or syrup is boiled — and other particulates and impurities. Sugar sand is not harmful if it is consumed, but it gives syrup an unpleasant graininess.

## FILTERING WITH ORLON

THE MOST COMMONLY USED MATERIAL to filter maple syrup is Orlon, a dense, synthetic acrylic material with a wool-like feel.

Orlon removes small unwanted particles and is used with a paper prefilter that removes larger unwanted particles. Orlon and paper prefilters can be purchased from most maple syrup supply dealers. When using an Orlon sheet or bag, always remember that the gentler you are with the bag or sheet, the longer it will last and the better job it will do. Don't twist or wring your Orlon while filtering or washing.

1. If using a new Orlon, dip it in boiling water for a few minutes before using to clean it and remove any residue that may flavor your syrup while you filter it.
2. Place your Orlon bag or sheet over the pail or pan that your syrup will be filtered into, then place one or two paper prefilters in the Orlon bag or sheet. Using two will allow you to filter out more of the larger particles so the syrup will go through the Orlon more easily.
3. Pour hot syrup (180°F [82°C] or higher) into the filter, and let it drain through. Put small batches, 1 or 2 gallons, of hot syrup through the filter at a time. Don't squeeze or force the syrup through the Orlon. This is meant to be a slow process. If you force the syrup through the Orlon, you will allow impurities to get through, and you may damage the Orlon.

If the syrup drains through too quickly, it may mean that the fibers in your Orlon have broken down and it's not catching the unwanted sediment. If this happens, it may be time to buy a new piece of Orlon.

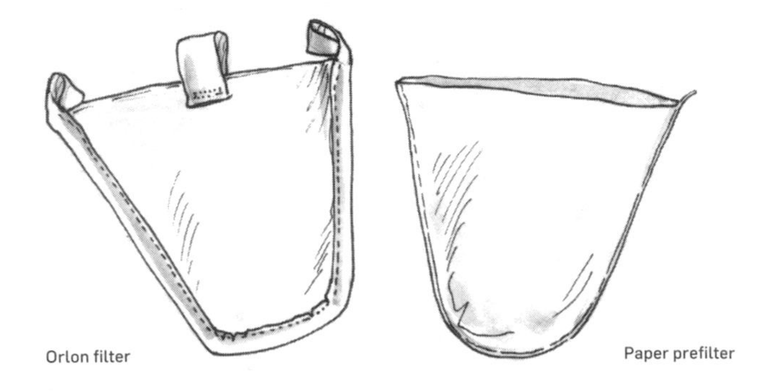


Filtering syrup can be a frustrating process, but Orlon will do a good job of removing unwanted sugar sand.

## Tips for Using Orlon

For best results follow these guidelines:

- **Never wash your Orlon with soap.** It will flavor your syrup. Instead, wash your Orlon only with hot water. You may turn the Orlon bag inside out to wash the inside, but be gentle.
- **Never twist or wring out your Orlon** or unduly pull or stretch it. This will make the material rupture, and it will do a poor job of filtering.
- **Always make sure your syrup is hot** — 180°F (82°C) or more — before filtering. Cold syrup will just sit in the filter and not strain through.
- **Slightly dampen the Orlon with hot water** before using. This will help your unfiltered syrup drain through. Take care not to get the Orlon too wet, as this may thin down your syrup.
- **Be patient!** Orlon will do a nice job, but it takes time and may need more than one pass.

### Troubleshooting with Orlon

One year you may have a hard time filtering your syrup and the next year it may go through more easily. This is because the amount of sugar sand — the sediment created when sap or syrup is boiled — varies greatly from year to year.

If after filtering you notice more sediment in your containers than you used to, it may be time to buy a new Orlon filter. Remember that Orlon is not perfect and will not always create crystal-clear syrup the way a filter press does. However, it works well for smaller producers.

Orlon will last a few years if taken care of properly. If rinsed carefully, the paper prefilters will last for a few batches. It is a good idea to have extra prefilters on hand.

---

## THE FILTER PRESS

A FILTER PRESS IS A DEVICE comprising a series of metal plates lined with filter papers and a pump that pushes the syrup through these plates. Any sediment in the syrup is left behind in the filter papers. Use a filter aid (diatomaceous earth) to help the press effectively remove sediment from the syrup. You will use less filter aid to move your syrup through the press at the beginning of the season than toward the end of the season — as the season goes on, you will make lower-quality syrup because of the bacterial growth in your tapholes and on your equipment.

There are different types and sizes of filter presses made by many manufacturers. The goal when using a filter press is to move the maximum amount of syrup through the pump

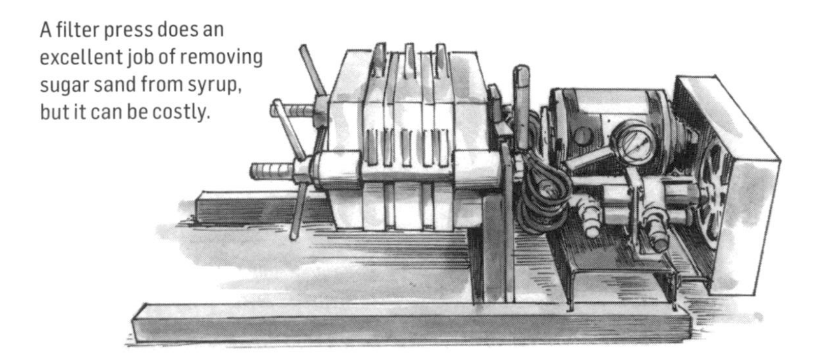
A filter press does an excellent job of removing sugar sand from syrup, but it can be costly.

according to the manufacturer's specifications. Generally, a full-bank pump will clean 40 gallons of syrup at one time. A short bank will clean 15 gallons, and a hand pump will clean 10 gallons. Check with the manufacturer of your pump to determine its actual size, how much syrup it can clean at one time, and how it should be assembled. Each manufacturer will have specifications unique to its filter press.

The pump on a filter press is motor driven, pneumatically driven, or manually driven. Motor-driven pumps are generally belt driven, and pneumatic-driven pumps use an air compressor. Manually driven pumps use a hand pump to push the syrup through the press plates. Hand-pump presses are less expensive than the other options, but they take more work.

Even a small producer can use a filter press. There are small, reasonably priced presses available that will work with as little as 5 gallons of syrup. Even the smallest filter presses do as good a job as the big filter presses, and they can make your syrup look as clear as if it were packaged professionally.

## How to Use a Filter Press

1. Bring your syrup to a boil. Test your syrup with a hydrometer or refractometer to make sure the density is correct, then remove your syrup from the heat. Add filter aid, and mix well.
2. Heat the syrup and filter aid to at least 200°F (93°C). Place the inlet hose from the filter press into the syrup pan.
3. Turn on the pump. Make sure the outlet hose for clean syrup is feeding back into the pan that you're pulling unfiltered syrup from. This is called recirculating. Do this until the plates and frame are warm to the touch.
4. Put a sample of clean syrup into a glass bottle and hold it up to the light. The syrup should be completely clear and free of sediment.
5. If the syrup is clear, stop the filter press and move the outlet hose to the pan that you will bottle your syrup from. Pump the clean syrup through the filter press into the canning pan. At this point make sure the syrup is up to proper canning temperature, between 180 and 200°F (82 and 93°C). If your syrup is not at least 180°F (82°C), you will need to heat it. *Do not boil* the syrup — boiling it will create more sugar sand, and you will have to refilter the syrup. When your syrup is at the proper bottling temperature, you can bottle it.

If the syrup is very cloudy, it isn't clean. This may be due to a broken filter paper, incorrectly assembled plates, or an upside-down plate. You'll have to take your press apart to find the problem, reassemble it properly, then try filtering it again, beginning with recirculation (step 3).

## How to Clean a Filter Press

When you have finished filtering a batch of syrup, or your press is starting to build more than 60 psi of pressure (you can see this by watching the pressure gauge), it's time to clean the press and change the filter papers.

1. Put the ends of both the outlet hose and the inlet hose into a pail of clean water of any temperature. Depending on the size of the press, you will use between 2 and 5 gallons of water. Recirculate the water through the press for 3 to 5 minutes. When finished recirculating, drain out the water from the hoses. You can save this water, boil it, and turn it back into syrup.
2. Loosen your filter plates, and pull them out. They will be filled with sediment and filter aid; try to keep all the sediment and filter aid in the cake frames (the hollow plates). Hold the cake frames over a trash can, and push the sediment into the garbage. If your press is small enough that the plates fit in a bucket, rinse them in clean water to remove additional sediment. If you have

a large press, do your best to wipe it clean before putting it back together.

3. If you're not going to use your press again for a week or so, run clean water through it, without papers between the frames, to make sure all syrup is removed before it can crystallize. If you are going to use your press again soon, put filter papers between the plates, and you're ready to go again.

Filtering syrup is often a lengthy and cumbersome process. Using a filter press can make the process easier. These tips should make running the press simpler — or at least less frustrating.

## BOTTLING SYRUP FOR THE HOME

ONCE YOUR SYRUP HAS BEEN FILTERED, you will want to put it in containers so it can be stored. If bottled properly, unopened syrup will never spoil. Properly bottled syrup can be stored at room temperature until it is opened, but once opened, syrup must be refrigerated.

Heat your syrup to between 180 and 205°F (82 and 96°C) before putting it in glass or plastic bottles. This will create enough heat to sterilize the bottle and seal the cap. If your syrup boils, it will form more sugar sand, and you'll have to refilter it. Check your syrup often with a thermometer.

## From Pan to Bottle

You can ladle your syrup from your pan into a funnel to get your syrup into the container. Or you can purchase a bottling unit from a syrup supply dealer. A bottling unit typically consists of a propane burner, a 16-by-16-inch (or similar size) pan that has a valve toward the bottom and a screen that fits on the top of the pan. These units work well for filtering with Orlon.

To use a bottling unit, simply pour hot syrup into the pan through an Orlon that sits on top of the screen. Turn the burner on to keep your syrup at the proper bottling temperature. Place your container under the valve, and open the valve to fill the container.

## Proper Cleanliness

If your syrup is hot (at least 180°F [82°C]), you do not need to pre-sterilize new containers or caps. Dipping plastic containers, plastic caps, or metal caps with plastic liners in boiling water will compromise their integrity, and they may fail when used.

If you are reusing glass containers, make sure they are clean. The best way to do this is to dip them in boiling water. Do not use soap; the residue left behind will flavor your syrup. You will also need to purchase new caps, as the old caps will likely not reseal. It is best not to reuse plastic containers because they are very difficult to clean. They may look clean, but plastic is porous and can hold bacteria that cannot be seen. Old plastic syrup containers should be recycled.

## BOTTLING AND STORING TIPS

**Sugar sand.** If sugar sand settles to the bottom of the bottles after you have finished packaging, fear not. Sugar sands are the naturally occurring particulates in maple syrup. They are not harmful if ingested; they just make your syrup a little gritty. Most of the sugar sands will stay at the bottom, and you can pour the clear syrup off the top.

**Crystals.** If large rock candy–like crystals appear in the containers, that may mean your syrup is too thick. Syrup stabilizes in liquid form somewhere between 66 Brix (32 Baume) and 68 Brix (38 Baume). If it is too thick when it is packaged, sugars in the syrup will drop from the liquid and form crystals until the liquid gets to the right density. These crystals will not develop right away, as it takes time for the syrup to stabilize in the container. To dissolve the crystals, you can add a small amount of water, then heat your syrup. Or when the syrup is gone, eat the crystals as you would candy. Generally, lighter-colored syrup crystallizes at a lower sugar level than darker-colored syrup.

**Storing thin syrup.** If you choose not to finish your syrup to 66 Brix and intend to keep the thin syrup for yourself, you will need to refrigerate it because thin syrup will spoil. It is important to note that it is illegal to sell thin syrup labeled as pure maple syrup.

**Storing finished syrup.** Syrup that is cooked properly and finished at 66 to 67 Brix or 32 Baume can be stored, unopened, for years when properly bottled, but it probably won't stick around that long! Once a bottle of syrup is opened, it should be stored in the refrigerator to prevent the growth of mold. Mold will appear as a powdery skim floating on the surface of the syrup. As with cheese, you can remove the mold and eat the rest of the syrup. If you'd prefer, you can boil the syrup to make sure any mold growth is eradicated, but this will create more sugar sand. Opened, finished syrup will stay good in the refrigerator indefinitely.

There are a variety of containers made specifically for packaging maple syrup, but you can store syrup in any well-sealed, clean container.

We have some antique syrup on display in our sugarhouse. Packaged between 1930 and 1950, the syrup in these containers has become darker in color with time. We have not opened the containers to taste the syrup, but there is no mold growth.

## *Old-Fashioned Tins*

Years ago it was a popular practice to package syrup in unlined tin containers. The problem with storing syrup in tin is that if it remains in the tin for longer than 6 months, the syrup will take on a "tinny" taste. In recent years, researchers found that some of the materials in the tin containers leached into the syrup, causing the tins to be recalled.

The tins available for sale today are lined with a safe, food-friendly material to prevent this leaching. The newer tins should have a better shelf life than old-style tins, but even so, glass is still the best way to store syrup.

CHAPTER 6

# COOKING ON AN EVAPORATOR

So you love making maple syrup. Your friends and family love that you make maple syrup. You realize that you need to make more syrup to satisfy the requests for your delicious maple syrup. If this is the case, you may want to think about investing in an evaporator, which is simply a cooking pan or a set of cooking pans placed over an arch (sometimes referred to as a firebox).

There are three reasons to think about cooking on an evaporator: You're tapping more trees, you need to cook faster, or you would like to cook more efficiently. If you can't invest any additional time in cooking sap, the quality of your syrup is suffering, or the syrup comes out very dark because it is over heat for too long, cooking on an evaporator may remedy these issues.

## WHAT IS AN EVAPORATOR?

An evaporator basically comprises stainless steel rectangular pans over a heat source called an arch. Most arches are built to accommodate wood or fuel oil, but they can also be heated with natural gas, propane, or high-pressure steam. Dry hardwood works best, as it burns very hot. Evaporator add-ons, such as reverse-osmosis machines and preheaters (see Additional Components, page 61), will reduce the amount of fuel used during a syrup season.

There are two basic types of evaporators: flat-pan evaporators and flue-pan evaporators. Flat-pan evaporators utilize a pan or pans that have flat bottoms. Flue-pan evaporators use a pan that has corrugated flues above or below the bottom of the pan.

### How Much Fuel?

*To make sure you have enough fuel to finish all your sap, use the following as a guide:*

- *1 cord of wood for every 100 taps*
- *1 gallon of fuel oil for every 10 taps*

## Flat-Pan Evaporators

Flat-pan evaporators are generally used by hobbyists who are cooking for fewer than 150 taps. Flat pans come with or without dividers; dividers allow the sap to flow through the pan in channels. Whether you cook in a pan with or without dividers, it is difficult to completely finish your syrup on a flat-pan evaporator. Most syrup makers who use flat-pan evaporators cook their sap until it is almost finished, then move the "thin" syrup into a smaller pan and finish it over a heat source that is easy to control, such as a propane burner.

Flat-pan evaporators are generally less expensive than flue-pan evaporators, but they still cook sap quite efficiently. Many manufacturers make such units, and they can be purchased from a maple syrup supply dealer. Ask a dealer for assistance when determining which style and size is right for your operation.

## Flue-Pan Evaporators

Flue-pan evaporators include two pans — one sitting on top of the back half of the arch (called the flue pan) and one sitting on top of the front half of the arch (called the syrup pan). The back of the evaporator is where the smokestack comes out of the arch. The wood or fuel goes into the arch at the front of the evaporator.

Sap flows continuously from the point of entry, where sap comes into the pan, to the point of draw off, where finished syrup leaves the pan. Cooler and less dense sap stays toward the back as the hotter and denser syrup travels to the front. As water evaporates from the sap, the sugar content increases, and the sap flows through the evaporator, until it is drawn off the front of the syrup pan as pure maple syrup.

**Raised or dropped flue pans.** The syrup pan is flat and has dividers. The back flue pan has either raised or dropped corrugated channels running through it. As sap enters the flue pan from a holding tank, a valve on the tank is open, and sap slowly and continuously drizzles into the pan. The channels running though the flue pan add surface area to the pan and will thus decrease cooking time.

In a raised flue pan, the channels are flush with the bottom of the pan and rise up into the pan. In a drop flue pan,

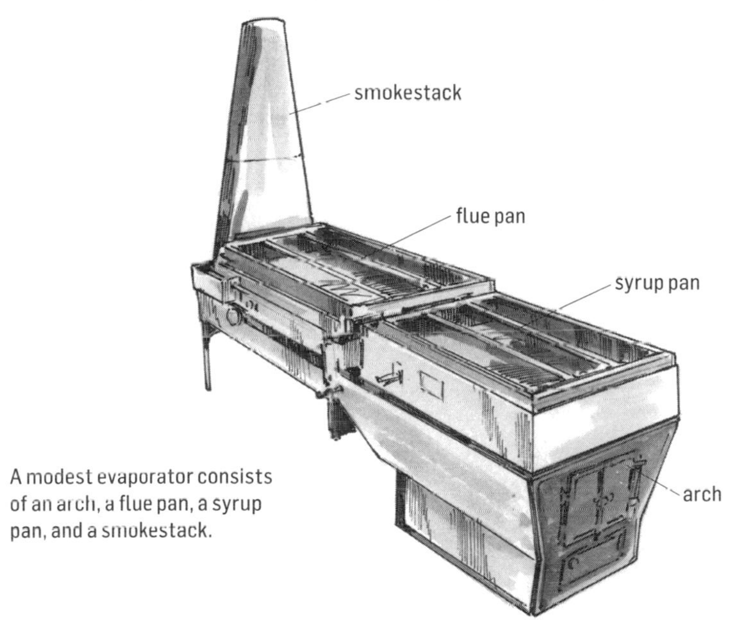

A modest evaporator consists of an arch, a flue pan, a syrup pan, and a smokestack.

the channels drop down below the bottom of the pan into the arch. A raised flue has two float boxes, and a drop flue has one float box; this is a regulating tool that controls the level of sap in your pan, keeping the level consistent during cooking. Floats utilize a valve system to regulate this level automatically. A float box consists of the box, the float, and the mechanism that opens and closes a valve.

Because it has two float boxes, a raised-flue evaporator allows the sap in your flue pan and the sap in your syrup pan to be at two different levels. Because of the difference in levels, sap can run more quickly from the flue pan into the syrup pan if necessary. This may come in handy if the level in your syrup pan is suddenly so low that it is in danger of going dry and burning. However, because it has two float boxes rather than

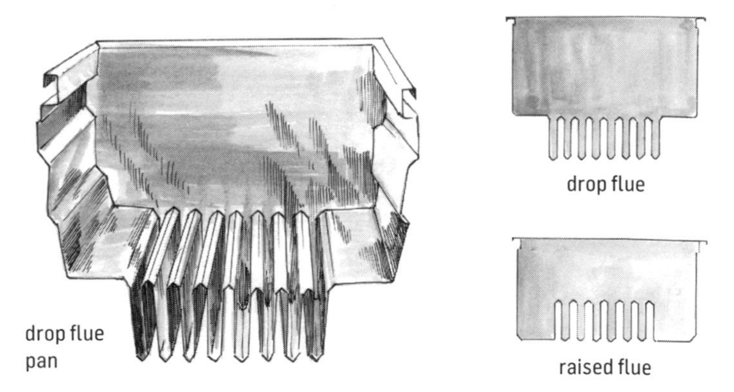


The flues in a raised flue pan rise up above the evaporator's arch. The flues in a drop flue pan drop down into the evaporator's arch.

one, a raised-flue evaporator has more moving parts that could potentially need replacing.

Functionally, those are the main differences between a raised- and a dropped-flue pan. The two pan types, if equal in size and number of channels, should cook off the same amount of sap per hour.

## Additional Components

You can purchase a number of components to customize your evaporator. Preheaters, steam hoods, and reverse-osmosis machines are just a few.

- A **preheater** heats and evaporates water from sap before it enters the flue pan. This prevents cold sap from coming into the flue pan and slowing down the evaporation process.
- A **steam hood** is a unit that hangs over the top of the evaporator and directs steam out of the cookhouse.
- A **reverse-osmosis machine** uses pressure and a semipermeable membrane to separate water from the sugar particles in sap. Reverse osmosis can remove up to 80 percent of the water before the sap is added to a flue pan. This cuts down cooking time considerably.

Evaporator add-ons can appear to be costly, but if you need to decrease your cooking time, they can be well worth the additional cost. A maple syrup supply dealer can help you calculate the long-term cost savings.

## PURCHASING AN EVAPORATOR

MOST EVAPORATORS ARE SPECIAL ORDERED well before the syrup season begins. You need to do some setup work to prepare for an evaporator, which you may not want to do in the snow and the cold. To avoid this, purchase your evaporator several months prior to syrup season.

Larger manufacturers begin building evaporators in the late spring and early summer. This is a good time to decide what you want and to place an order. Purchasing an evaporator during the summer months will ensure that you'll get exactly what you want. You'll receive your evaporator well before syrup season, and you'll have ample time to set it up before the sap starts to run. However, if you're fairly flexible and intend to buy a relatively basic model, a dealer near you may have something in stock that will suit your needs.

### *The Advantage of an Evaporator*

The longer sap is exposed to heat, the darker in color and stronger in flavor it gets, possibly taking the syrup from one grade to the next. An evaporator decreases the time that sap is exposed to heat because you can continually draw finished syrup off the syrup pan. Therefore an evaporator can potentially make better-quality syrup than a flat pan. It is possible to make nice light syrup on a flat pan; it is just easier to do it with an evaporator.

## SIZING AN EVAPORATOR

When purchasing an evaporator, plan for a middle-of-the-road sap run, not the best- or worst-case scenarios. This will ensure that your evaporator is not under- or oversize for your production.

Evaporators come in various sizes. The best size for you will depend on a number of factors:

- The number of taps or amount of sap that you will be processing
- Whether you are collecting via gravity or vacuum
- The number of hours per day you will be able to spend cooking your sap (the more time you have to spend cooking your sap, the smaller the evaporator you can buy)
- Whether you will ever increase the number of taps you put out or the amount of sap that you process (by buying sap from other people, for instance)

**With a gravity system.** On an average day, if you are collecting by gravity (either in pails or a gravity-flow tubing system), you can plan for 1 gallon of sap per tap. So if you have 100 taps and 10 hours to complete your cooking, you will need an evaporator that will boil off 10 gallons of sap per hour.

**With vacuum tubing.** If you are collecting with a vacuum tubing system, you can plan for an average of 1.5 gallons of sap per tap per day. Again, if you have 100 taps and 10 hours to complete your cooking, you will need an evaporator that will boil off 15 gallons of sap per hour.

### Sizing Your Pans

The evaporation rate is about 1 gallon per square foot of surface area. So to cook off 10 gallons of sap per hour, you'll need a pan with 10 square feet of surface area. A 2-foot-by-5-foot or a 3-foot-by-3.5-foot flat pan will give you approximately the right square footage. In a flue pan, the channels will add surface area, giving a smaller pan a lot more square footage.

To maintain this cooking rate, you'll also need to maintain a good hot fire underneath your pan. For instance, it's a good practice to fire very consistently, every 10 to 15 minutes, with very dry hardwood. Some producers set a timer so they can maintain a consistent firing schedule.

Consult a reputable maple syrup equipment dealer for advice on the appropriate size pan or evaporator for your operation.

---

## INSTALLING YOUR EVAPORATOR

THE ARCH OF AN EVAPORATOR needs to be insulated for maximum efficiency. When cooking with wood, the arch is generally insulated with firebrick. When cooking with oil or gas, the arch is insulated with a ceramic blanket or firebrick.

Oil- or gas-fired arches often come insulated, but wood-fired arches do not. Once firebrick is installed, the evaporator becomes extremely heavy. The evaporator would be very difficult to move if it were bricked prior to putting it in place. When purchasing an evaporator, make sure your dealer gives you bricking instructions, so you can install the brick once you have placed your evaporator in the desired location.

## Foundation

It is best to place your evaporator on a concrete floor. If this is not possible or practical, you may place it on patio blocks or a level layer of gravel. The foundation for your evaporator needs to be level and must remain so as long as your evaporator sits there. You should not place an evaporator on bare ground. The freeze-thaw cycle will cause the evaporator to sink into the soil and become unlevel. If your evaporator is not level, the sap will not flow evenly through the pans. If the sap is uneven in the pans, a shallow spot could develop during cooking, and the pan could scorch or burn.

In a wood-fired evaporator, the arch is almost always open on the bottom, under the grates. Hot ash and embers will fall through to the ground. You certainly don't want to install your evaporator on a wood floor!

## Smokestack

Once your evaporator is in position, make sure the smokestack lines up with the hole in your roof before installing the fire-brick. That way, you can adjust the evaporator's position before it is too heavy to move easily. If your arch is wood fired, your smokestack should be twice as tall as your pans are long. If your arch is oil fired, your stack can be shorter. With either type of arch, your stack must clear the peak of the roof of your cookhouse by 3 feet to maintain a good draft.

CHAPTER 7

# COLLECTING SAP WITH TUBING

Producers who want or need to reduce the amount of manual labor required to collect sap often install tubing systems. These consist of flexible plastic tubes and the fittings that connect them. Sap tubing is specially made for the syrup industry of plastic that is formulated to the proper flexibility, elasticity, and memory for hanging in the woods at varied spring temperatures. The special plastic allows some sunlight to pass through it, but not enough to heat the sap. This is important, because the warmer the lines get, the faster bacteria grows. The more bacteria in the sap, the darker your syrup will be.

Tubing and tubing fittings can work very well for a producer if used properly. A tubing system can be installed in a variety of ways, depending on the slope and topography of your woods. If you are unsure of how to install a tubing system, it is best to consult an installer or a dealer for assistance. Tour a sugar bush similar to your own that uses a tubing system for sap collection. Find some good resources, and have a plan before you begin.

### Not Just Any Tubing

*Your hardware store may sell tubing that would work to collect sap, but it will not perform as well as sap tubing. Tubing is installed to increase productivity and decrease manual labor. If the tubing you use is not specific to the maple industry, it will compromise your productivity.*

**Shopping List for a 500-Tap Tubing System**

- 500 plastic tubing spouts (we recommend 5⁄16-inch)
- 400 tubing tees that match the spouts you choose (the plug on the tee needs to be compatible with your spouts)
- 100 end fittings (we recommend Leader Evaporator's End Ring)
- 100 hooked connectors
- 100 1-inch mainline inlets

*List continues on following page*

- 3 rolls of 5⁄16-inch drop line tubing
- 20 rolls of 5⁄16-inch lateral line tubing
- 2,000 feet of 1-inch mainline tubing
- 2,000 feet of mainline wire (12.5-gauge high-tensile wire)
- 2,000 5-inch wire ties
- 6 7⁄16-inch lag bolt insulators or lag I-bolts
- 10 wire locks or Gripples for 12.5-gauge wire
- 1 500-foot role of side-tie wire (not high tensile)
- 9 1-inch pipe thread to hose adaptors (plastic is fine)
- 6 1-inch valves (plastic with pipe threads)
- 3 1-inch hose connectors
- 15 hose clamps for 1-inch hose
- Collection tank to gather sap (If this is your main collection tank, and all of your sap is coming to one spot, you will want at least a 1,000-gallon tank; stainless steel is best, but plastic is fine.)
- Mobile tank to transfer sap from the woods to your sugarhouse, if needed

## Feeding the Evaporator

*Put a tee and valve at the tank, and have one line going straight up and another line feeding down to evaporator, then one going straight up. This will work as a sight gauge to tell you how much sap is in your upper tank, so it needs to be 12 inches higher than your tank.*

- 1 additional 1,000-gallon storage tank at your sugarhouse
- Transfer pump to raise your sap to a small elevated tank that feeds your evaporator
- Small elevated tank (feed tank), plus any necessary plumbing that goes with the feed tank
- Plumbing to feed your evaporator (use clear hoses on elevated tanks to be able to see the sap)
- Evaporator that will boil off at least 50 to 60 gallons of sap per hour, such as a standard 30-inch-by-10-foot evaporator with a flue pan and a wood-fired arch (or consult a maple syrup supply dealer for other options)
- 1 thermometer that can work with syrup pan and bottling pan

**Additional Items for Vacuum Flow**

- 1 8 CFM (cubic feet per minute) vacuum pump (gas or electric)
- 1 small releaser (something that can handle 500 taps)
- 1 moisture trap to protect your vacuum pump (goes between releaser and vacuum pump; we recommend purchasing a vacuum pump with the moisture trap built in)
- Smaller, lower-profile tank for your releaser to empty into to keep your system going if there are height restrictions
- 4 vacuum gauges to place throughout your system to help monitor leaks

*List continues on following page*

**More Extras**

- Automatic draw-off
- Draw-off tank
- Small filter press, filter papers, and filter aid
- Steam hoods
- Syrup hydrometer and cup
- Bottling pan with a heat source and thermometer
- Valve for bottling pan
- Containers, covers, and labels

## TWO TYPES OF TUBING SYSTEMS

YOU CAN USE EITHER A GRAVITY-FLOW or a vacuum-flow tubing system. A gravity-flow system uses natural gravity to move sap from the trees to a holding facility. A vacuum system uses a vacuum pump — either gas or electric — to move sap from the trees to a holding facility. An electric pump will require electricity in the pump house or at the holding facility; a gas-powered pump will run on a generator or a gas-powered motor. A vacuum pump works by removing air from the tubing, thus creating lower pressure in the lines and allowing sap to flow quickly from the trees. This is the same principle that causes sap to run naturally into a gravity-flow tubing system or into buckets. The vacuum system just enhances the low pressure that nature creates.

A properly installed and maintained vacuum system can more than double sap production. Many producers find that the added sap production is worth the additional cost of installing a vacuum system.

## Installation Difference

Gravity-flow and vacuum-flow tubing systems are installed similarly, with one exception: In a vacuum system there is a strict guideline for how many taps you can put on each line of tubing; there are no such guidelines in a gravity system. However, if you are putting in a gravity system initially but think that you may add vacuum to your lines in the future, you should plan for that during your initial installation. This will make it much easier to add vacuum in the future.

The tubing installation techniques that follow allow for a vacuum system, either now or in the future. If you intend to install a tubing system that is strictly gravity flow and have no intention of ever adding vacuum, you may want to modify your installation somewhat from the instructions we provide. You'll be able to add a few more taps to your lateral lines. However, whether you install a gravity-flow or a vacuum-flow tubing system, we highly recommend consulting a maple syrup supply dealer or a tubing system installer before purchasing your supplies and installing your tubing. Each woods varies vastly from the next, and it's best to receive guidance tailored to your situation to ensure that you produce the highest sap output possible.

## INSTALLATION TIPS

REGARDLESS OF THE TYPE OF TUBING SYSTEM you are using, tubing should be installed so it continuously slopes downhill or down grade. There are a few additional guidelines to keep in mind when installing either type of system:

- **Fittings can restrict sap flow.** Keep your system as simple as possible. We will make suggestions for fittings throughout the installation guidelines.
- **A tubing system takes a considerable amount of upkeep.** Most producers who gather sap with tubing walk a portion of their woods each day during the season to check for leaks in the tubing, vacuum leaks, or tree limbs or other debris on the lines.
- **Tubing systems need to be inspected each year.** This should be done well before the syrup season begins, in case any major repairs need to be made.
- **Taps need to be pulled and tubing needs to be cleaned thoroughly** at the end of each syrup season.
- **Unlike metal, plastic will deteriorate.** Portions of your system, such as the spouts or spout extenders, will need to be replaced each year. Your entire system will need to be replaced every 10 to 15 years.

## LAYING IT OUT

Both gravity-flow and vacuum-flow tubing systems consist of drop lines, lateral lines, and mainlines, as well as tubing spouts and tubing fittings. Drop lines carry sap from the tree to the lateral lines. Lateral lines carry sap from drop lines to mainlines. Mainlines are the larger tubes that carry the sap to the collection site. Tubing spouts are tapped into the tree and fit into the drop lines, and tubing fittings connect the entire system.

1. **Begin by surveying your woods.** Your lateral lines should run on the steepest slope of the hill, and your mainlines on the secondary slope. This is especially important with very steep grades.
2. **Decide where your mainlines will run.** Mainlines should be no more than 1,000 feet long and as straight as possible. If your mainline needs to be longer than 1,000 feet, consult a maple syrup supply dealer for recommendations.

   Mainlines should be hung roughly 100 feet apart across the side of a hill and should maintain a 2 to 5 percent slope toward your pump house. This configuration will allow for lateral lines on the steepest part of the hill using as much slope as is available.

3. **Tie flagging tape around the trees to mark where your mainlines will go.** Each time you do so, take a reading with a sight level to determine where to mark the next tree. After flagging, cut a brush path about 5 feet on either side of the mainline area so you can access the mainline with an all-terrain vehicle.
4. **Hang the wire that will hold up your mainline.** Use a high-tensile wire that is either 9 or 12.5 gauge. Anchor it into a hardwood tree, preferably an oak or a maple, at each end of your woods. Begin at one end by attaching the wire to your anchor tree with a heavy-duty 7⁄16-inch I-bolt or a lag bolt insulator. Use a Gripple, or a one-way slide fastener, so the wire cannot slip backward and cause slack in the line.
5. **Pull the wire through the woods to the anchor tree on the opposite end.** Watch closely to make sure you are not pulling the wire against trees. Also pay attention to trees adjacent to the mainline that can be used for attaching side-tie wires to help keep the mainline tight. Pull the wire as tight as you can. In many cases, such as on a long mainline, the wire will still be lying on the ground after you pull it as tight as you can. It is just too physically difficult to pull it completely tight. Side tying will correct this. Anchor this end of the wire the same way you anchored the opposite end.

Anchor your mainline wire into a hardwood tree with a heavy-duty 7⁄16-inch I-bolt or a lag bolt insulator. Use a Gripple so the wire won't slide backward.

## *Keep the Slope under 5 Percent*

It is important to keep the grade of the mainline's slope at less than 5 percent; if it is steeper than 5 percent, sap can tumble through the tubing. If sap is tumbling, it will block the line and impede the transfer of vacuum. If you intend to install a vacuum system and cannot avoid a slope greater than 5 percent on a mainline, consult a maple syrup supply dealer for options. If you're collecting sap via gravity flow, tumbling sap is not an impediment and may actually create a slight vacuum. If you never intend to add vacuum, the grade of your mainline can be greater than 5 percent.

## INSTALLING MAINLINE

MAINLINES ARE PLASTIC TUBES, usually colored blue, that generally measure between ½ and 1½ inches in diameter. Mainline fittings are sold in various sizes to accommodate the different-size tubing.

Size your mainline tubing based on the number of taps you have:

- **For up to 50 taps,** use ½-inch mainline
- **For up to 250 taps,** use ¾-inch mainline
- **For up to 500 taps,** use 1-inch mainline
- **For up to 1,000 taps,** use 1¼-inch mainline
- **For up to 2,000 taps,** use 1½-inch mainline

Install mainline tubing by tying it to the mainline wire with wire ties.

Using wire ties, tie your mainline tubing to the wire that you already placed in your woods. Attach a wire tie every 6 to 10 feet. After your system is in place, you can go back and attach a wire tie every 12 inches. One end of the mainline will hook into your holding tank or releaser. The other end will be plugged, or it will have a valve on it that can be opened for cleaning purposes.

## Tying Side Wires

Side wires are used to keep the mainline wire tight. Use a lighter (not high tensile), 14- to 16-gauge wire for your side ties.

1. Run your side-tie wire through a piece of 5⁄16-inch tubing. This keeps the wire from digging into the tree.
2. Fasten the side wire to the mainline wire.
3. Push against the mainline wire with your hip while pulling the side wire around a side tree. The side wire now makes a big U around the side tree.
4. Fasten the other end of the side wire to the mainline wire. Make sure the two ties are at least 12 inches apart. If the ties slide together, they'll kink the mainline wire, and it will break.

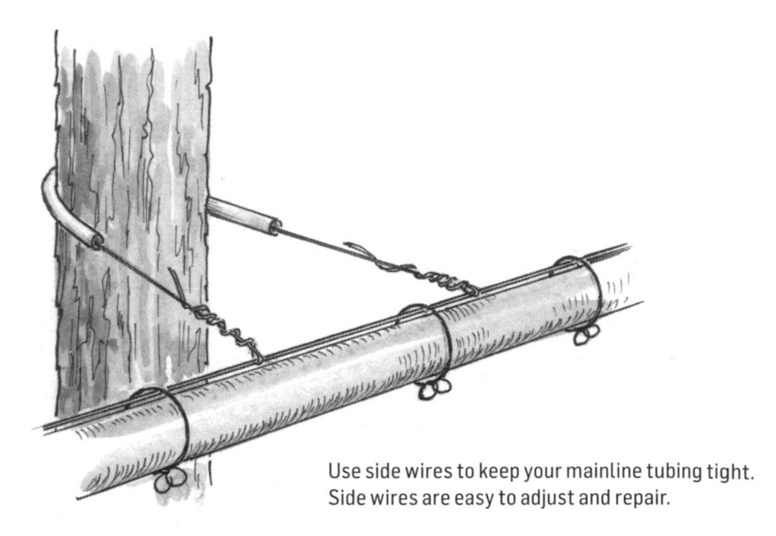

Use side wires to keep your mainline tubing tight. Side wires are easy to adjust and repair.

Continue tying side wires until the mainline wire is tight. If you set both ends of the mainline correctly, you can go back and adjust the height of the mainline later. This is an advantage to using side ties instead of stakes. Another advantage is ease of repairs. If a tree falls on your mainline, a side-tie wire will break, and that part of the line will fall to the ground. It is designed to do this. It is much easier and faster to repair a side-tie wire than to replace your entire mainline.

### Gauging Gauge

*The lighter the gauge, the higher the number.*

## INSTALLING LATERAL LINES

ONCE YOUR MAINLINES ARE HANGING in the woods, you can begin hanging your lateral lines. Lateral lines connect the drop lines from the trees to the mainlines. Lateral lines are also plastic tubes, usually colored blue, though you can find them in many other colors. They are always 5⁄16-inch tubes. All tubing spouts and lateral line fittings are sized to fit 5⁄16-inch lateral line.

Tubing installers have a saying for the number of taps you should put on each lateral line. "Strive for five, no more than eight, less is best." Try to put only five taps on every 100-foot length of lateral line. Using these guidelines, you'll average between 20 and 25 feet of lateral line tubing for every tap in your woods. There is always an occasion to add an extra tap to a lateral line, however. These aren't hard-and-fast rules, but the more you can stick to these guidelines, the better your system will perform.

1. While standing perpendicular to the mainline with your arms in front of you, shoulder width apart, find the tappable trees between your arms. Those are the trees you will tap with your lateral lines. You may want to mark these trees with tree-marking paint or tape. Avoid big zigzags to keep your line about 100 feet long or less.

2. Leave the roll of lateral line on the ground at the mainline and walk out to the last tree on your line, pulling an end while you walk and uncoiling the tubing as you go. Install an end fitting to the last tree on the lateral line. The tubing makes a loop around the tree and slides through the fitting to hold the line in place.
3. Walk back to your mainline, pulling your lateral line tight as you go. You may want to walk a slight weave down your marked line, using the trees for tension against the tubing.
4. Cut the lateral line a few inches short of reaching the mainline. You want to make sure there is a lot of tension on the line so it doesn't sag. Install a hooked connector into the lateral line tubing. Hook the connector over the mainline wire.
5. Install a saddle manifold. This is a fitting that wraps around the mainline and cinches like a saddle does on a horse. A variety of different manufacturers make saddle manifolds, so they come in different sizes. Begin by drilling a hole in the top of the mainline according to the manifold manufacturer's specifications. The saddle fitting goes into the hole, and a rubber gasket on the underside of the saddle makes a seal on the mainline tubing when the saddle is cinched tight.

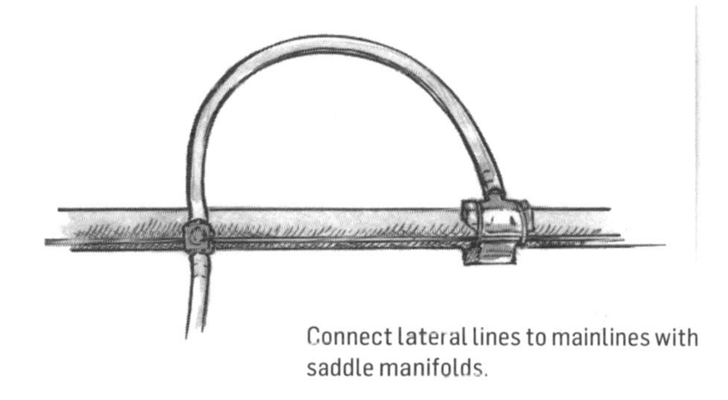

Connect lateral lines to mainlines with saddle manifolds.

6. Push one end of a short line of tubing (about 12 inches) onto the fitting on the top side of the saddle manifold. Push the other end of the tubing onto the unused end of the hooked connector. This will connect the mainline to the lateral line.

Hang a lateral line wherever necessary. It's important that your lateral lines be as straight as possible and no more than 100 feet long.

## How Much Tubing Should I Buy?

*As a general rule of thumb:*

*Number of taps × 20 = Number of feet of lateral line tubing to purchase*

*Number of taps × 30 = Number of inches of drop line tubing to purchase*

## INSTALLING DROP LINES

DROP LINES CONNECT THE TREE to the lateral line. They consist of 30 to 36 inches of 5⁄16-inch drop line tubing (usually blue), a plastic tubing spout, and a tubing tee. Drop line tubing is made of more flexible plastic than lateral line tubing and grips extra tight around fittings. Good drop line tubing will not allow fittings to spin or pull out. Drop lines are the last things to be installed in the system.

Drop lines can be assembled indoors prior to installation. This is helpful because cold air can make tubing less pliable and harder to work with; building the drop lines inside can make the process easier.

Assemble drop lines by cutting as many 30- to 36-inch lengths of drop line tubing as you will need — one for each tap, minus the last tap on the end of each lateral line. There you will put a drop-line lateral end. Install a tubing spout into one end of the length of tubing and a tubing tee into the other end. When assembling and installing drop lines, keep these tips in mind:

- **Make one drop line for every tap** you'll put in your woods, except for the last tap on a lateral line. For this, build a drop-line lateral end.
- **The last drop on the lateral line will not have a tee;** it will have an end ring or some other lateral line end fitting that will create the end of the lateral line. Build a drop-line lateral end for every lateral line you have in the woods.

- **To prevent reverse pressures,** which will pull sap that is up to 28 inches away back into the tree, drop lines should be 30 to 36 inches long. A 30- to 36-inch line will also allow you to move your tap around the tree in subsequent years. A 36-inch drop line will reach more than halfway around most trees.
- **Use a tubing installation tool,** such as a deluxe fitting assembly tool, to cut into the lateral lines. A deluxe fitting assembly tool has two clamp vise grips, one on each end of the head of the tool. To install drop lines, open the tool and clamp each vise grip onto the lateral line where you will attach the drop line. There will be a 3- to 5-inch length of lateral line between the tool's clamps. Cut this section of tubing out of the lateral line with a tubing cutter or pruning shears.

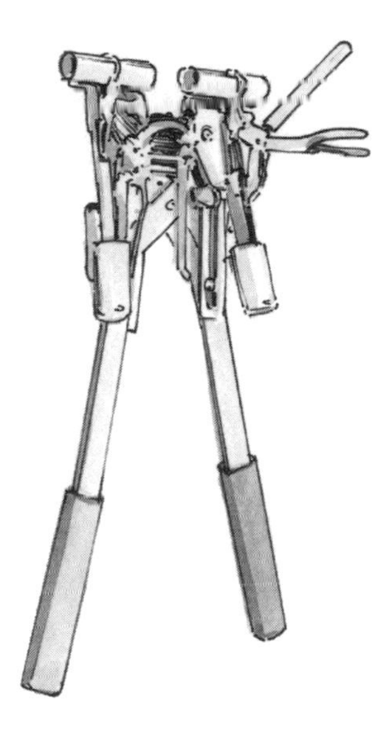

Use a tubing installation tool such as a deluxe fitting assembly tool to install drop lines.

- **Use a tee** to connect the drop lines into the lateral lines. Hold the tee in the space you cut from the lateral line. Squeeze the handles of the deluxe fitting assembly tool together, forcing the tubing onto the tee fitting until the tubing is all the way onto the tee. The tee will allow the sap running through the lateral line to go straight. It will also allow the sap from the drop line to run into the lateral line at a 90-degree angle.

Drop lines consist of drop line tubing, a plastic tubing spout, and a tubing tee.

## TUBING FITTINGS

TUBING AND TUBING FITTINGS can be purchased from a maple syrup supply dealer. There are only a few companies that manufacture them for the maple syrup industry. Sizes are mostly standardized, but occasionally one manufacturer's fitting will not work with another's. To maintain continuity throughout your tubing system and to make it easier to make repairs, it's a good idea to buy all your tubing and fittings from the same manufacturer.

Tubing spouts, like metal bucket spouts, come in two sizes: 5⁄16 inch or 7⁄16 inch. With tubing, 5⁄16-inch spouts are most commonly used. The end that goes into the tree is the sized end; the opposite end of the spout is always sized to fit 5⁄16-inch tubing.

Other basic tubing fittings include tees, connectors, saddles, and valves. Consult with a maple syrup supply dealer or a tubing installer to learn which fittings will work best with your system.

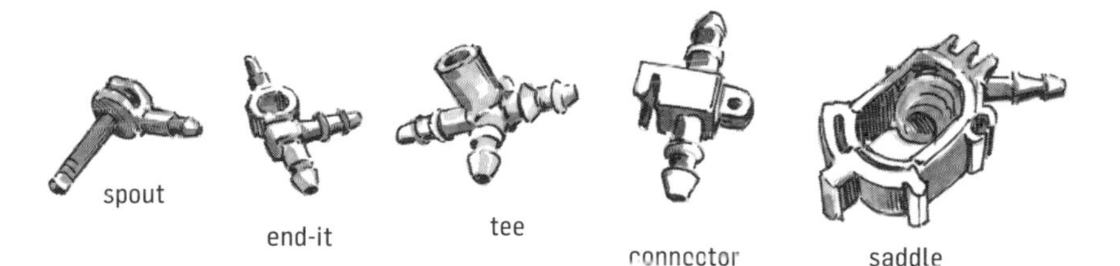

## PUMP HOUSE OR COLLECTION AREA

Once the tubing is installed in your woods, you are nearly ready to collect sap. You'll want to make sure your pump house or collection area is ready to go. Your pump house will hold a collection tank, a releaser, and a vacuum pump. If you have a gravity-flow system, you simply need a collection tank to hold the sap. Position the tank low enough, down grade, so your sap can flow into it easily. Consult a dealer to help you determine the size of the equipment you need for your operation.

For both gravity-flow and vacuum-flow systems, you need a transfer pump to move your sap to an elevated tank that will feed your evaporator. If you have to transport your sap to your cooking location, you will need a portable tank into which you can pump the sap from your collection tank and move it to your evaporator.

CHAPTER 8

# END-OF-SEASON CARE

There is no one signal that tells us when to stop collecting sap. When the temperature doesn't go below freezing at night and the days warm into the 60s (15–21°C), the syrup season is coming to an end. When the maple trees start to bud, the sap will become bitter. Syrup made from this sap will be off-flavored and taste "buddy." Some trees may have even stopped giving sap by this time because the tapholes may have begun to heal. This is how you know it's time to remove the spouts and clean up for the season.

## Orlon Care

When you are finished filtering your syrup, rinse your Orlon thoroughly with hot water. Do not wring it out; let it dry completely and store it in a dry place. If you find that your Orlon has become moldy when you come to use it again, dispose of it. *Do not* try to clean it. Even if the Orlon appears clean, the mold that is embedded in the material will give your syrup an off-flavor. Cleaners such as bleach or laundry detergent will also flavor your syrup.

Prefilter papers can be rinsed and reused a few times during the filtering process, but when the material begins to break down and become less stiff, throw them out. As with your Orlon, any unused prefilter papers should be stored in a dry place.

## Spouts, Pails, and Sap Saks

Pull the spouts out of the trees at the end of the season, using pliers, the claw end of a hammer, or a spout puller. *Do not* plug the hole in the tree. The tree will heal itself naturally, and plugging the hole can actually damage the tree.

Boil your spouts, and rinse the pails and covers with hot water. You do not need to use bleach or soap to clean your equipment. Some buckets may need to be lightly scrubbed with a plastic-bristled brush. If you used Sap Sak holders, take them apart, discard the sap bags, and rinse the holders with hot water.

Dry the spouts, pails, and Sap Sak holders thoroughly, stack them, and store them in a dry place. Store any unused sap bags in a sealed box so moisture and animals cannot get to them. Moisture may cause mildew to grow in the bags, and animals may chew on them.

### Formaldehyde Pellets

*Occasionally a customer asks about formaldehyde pellets. Years ago it was not uncommon for syrup producers to put formaldehyde pellets in tapholes to slow down the healing process, thereby increasing the length of the sap-gathering season. During the 1980s they became illegal; they are no longer available, as they are harmful to the trees and potentially to humans as well.*

## Syrup Pan and Hydrometer

Clean your syrup pan by letting it soak in rainwater. Unlike groundwater, fresh rainwater has no chemicals and will draw sugar sand deposits out of the pan. You can also spray your pan with a pressure washer.

In extreme cases use a special acid or pan cleaner available from most maple syrup supply dealers. These acids cannot be shipped, so if you are in need of this type of cleaner, you will have to see a dealer to acquire them.

Clean your hydrometer and cup with hot water. Carefully wrap the hydrometer, and store it in a place where it cannot fall or be knocked down.

## Tubing and Woods Care

During the off-season you'll want to remove damaged branches and dead trees from your wood. This is especially true if you have installed a tubing system, but you'll want to do this even if you're collecting sap in buckets. Dead trees and damaged branches have the potential to fall on your tubing. If you can eliminate most of them, you will lessen the repairs you'll have to make to your tubing. Also, when you are collecting with buckets, clear paths make it easier to work in the woods.

It is important to maintain the health of your sugar bush.

Inspecting and repairing your system during the off-season will minimize problems during the syrup season. Since weather and nature constantly cause disruptions in your system, it is a good idea to walk your woods regularly. This way you can find problems, such as branches on lines, and correct them as they come up. If you wait until just before the syrup season to inspect your system, you may have snow, cold temperatures, and a lot of repairs.

If possible, store all your maple syrup–making equipment in one place. It'll be easier to take inventory when you purchase additional equipment for next season. And when the next season arrives, all of your equipment will be in one place, clean and ready to go.

Now you have gone through the entire process of making pure maple syrup. Make yourself a stack of pancakes, pour on the results of your labor, and enjoy it!

CHAPTER 9

# SUGARHOUSE CONSIDERATIONS

To allow producers to sell syrup, many states and provinces require that they be licensed. As part of the licensing process, states and provinces may require that producers follow certain mandates. Frequently, one such requirement stipulates that the producer have a designated area for syrup production and another for packaging. For example, the area that you designate for packaging your syrup cannot also double as the family kitchen!

Because licensing requirements vary from state to state and province to province, it's best to check with your local inspector to learn what they are in your area. Many times the syrup inspector will be the dairy or agriculture inspector, but if you don't know who the syrup inspector is in your area, check with your state or provincial syrup association. Your association can tell you how to obtain a license and how to contact an inspector. The North American Maple Syrup Council (NAMSC) has compiled a list of state and provincial syrup associations (available on their website; see Resources).

### NAMSC

*The North American Maple Syrup Council (NAMSC) educates producers about the newest industry practices and maple syrup research. According to the organization's website (see Resources), NAMSC "brings together industry leaders and affiliated groups to share common interests, experience, and knowledge for the advancement and improvement of the maple syrup industry." The organization also educates consumers about the difference between pure and imitation maple syrup, and the nutritional benefits of pure maple syrup.*

## LOCATION

YOU MUST BE ABLE TO ACCESS YOUR SUGARHOUSE, which needs two things. One: The building needs electricity. Two: You need to be able to get sap there. If you find a location where both of these are convenient, you've lucked out! If you have a location that is convenient for only one, you'll need to make modifications so that both electricity and sap reach your building.

## IMPORTANT SUGARHOUSE ELEMENTS

SOME BASIC ELEMENTS that should be included in a sugar-house are a concrete floor with drains, hand-washing sinks, ventilation for steam, an evaporator smokestack, electricity and lighting, a separate storage area for fuel — whether it be wood or oil — and doors large enough to take equipment in and out. Here are other things to keep in mind.

**Clean surfaces.** We've found that inspectors like to see clean ceilings or covered evaporators to eliminate the chance that something can drop into an open syrup pan. Ceilings should be steel or wood that is painted or varnished. If the visible wood is exposed or unfinished, your pans should be covered. A steam hood will work.

**Room for an evaporator.** At minimum there should be enough room to accommodate the size of your evaporator plus an additional 4 feet of working space around the evaporator. The building will need to be high enough to accommodate the height of your evaporator, its cover, and any other add-ons. You will want some extra space for a work area. Consider how you will store the syrup you have made: in barrels, milk cans, or bulk tanks? Include storage space for finished syrup.

**Room for firing.** If you are using a wood-fired evaporator, you will need additional space in front for feeding wood into the arch. You may want to elevate your wood-fired evaporator or build a pit in front of the arch to stand in. Firing is much easier if you don't have to constantly bend over. Remember: You'll be firing consistently every 10 to 15 minutes.

**Room for visitors.** Once you've calculated the size of your evaporator and any additional space you might need, add some space for visitors. Family and neighbors will want to watch you make syrup. Keep safety in mind when you're planning for visitors, as there is some danger involved in syrup making because of the heat created by the evaporator.

## FUEL STORAGE

WHEN COOKING ON A WOOD-FIRED EVAPORATOR, it will take approximately one cord of wood for every 75 to 100 taps. A full cord of wood measures 4 feet wide, 4 feet high, and 8 feet long. Hardwoods like oak and maple work best, and they should be very dry. It is best to store your wood in a covered or enclosed structure.

If you are using fuel oil, the tank should be secure and easily accessed by fuel delivery trucks. Plan to store 1 gallon of fuel for every 10 taps. If your tank is not easily accessed, you'll have to plan for more storage.

### How Much Is a Cord of Wood?

*A cord is a unit of measure of volume of cut timber. A cord of wood occupies a volume of 128 cubic feet or an area that typically measures 4 feet wide, 4 feet high, and 8 feet long — or any arrangement that represents the same volume.*

CHAPTER 10

# GRADING AND SELLING YOUR MAPLE SYRUP

Once you've done all the hard work of collecting sap and making syrup, take a minute to marvel at the beautiful product you've made. Maple syrup is not only delicious; it's beautiful to look at, too! Note the color of your syrup. Is it a light golden hue or a dark amber? Taste your syrup. Does it have a mild, delicate maple taste or a strong, robust maple flavor? To place a grade on your syrup, you need to determine its color and flavor category.

## GRADING YOUR SYRUP

You must grade your maple syrup to place a proper value on it for sale purposes, though not all states and provinces require that the grade be shown on the product. In the United States, pure maple syrup is graded according to federal USDA regulations. The USDA uses five grades for syrup: Grade A Light Amber, Grade A Medium Amber, Grade A Dark Amber, Grade B, and Commercial Grade. Vermont, Ohio, New Hampshire, New York, and Maine have their own variations on these grades. In Quebec, which produces about 70 percent of the world's syrup supply, syrup is graded with the letters AA, A, B, and C.

The International Maple Syrup Institute (IMSI) has come up with a standardized system for grading pure maple syrup that will replace the current grading systems in the United States and Canada. The IMSI grading scale focuses more on flavor than color, whereas many other grading systems focus

### IMSI

*According to its website, the International Maple Syrup Institute (IMSI) is an organization that works "to promote the use of pure maple syrup and protect the integrity of the product while encouraging cooperation among all persons or groups involved in any aspect of the maple industry."*

mostly on color. Flavor can't be measured with a tool; you have to learn how to grade syrup by taste, which takes practice and trial and error. You can use a comparator to grade the color of syrup. A comparator, like a maple syrup grading kit, compares a sample of maple syrup to a standardized color sample.

IMSI suggests four classes of pure maple syrup:

- **Golden Maple Syrup** "has a light to more pronounced golden color and a delicate or mild taste."
- **Amber Maple Syrup** "has a light amber color and a rich or full-bodied taste."
- **Dark Amber Syrup** "has a dark color and a more robust or stronger taste than syrup in lighter color classes."
- **Very Dark Maple Syrup** "has a very strong taste. It is generally recommended for cooking purposes but some consumers may prefer it for table use."

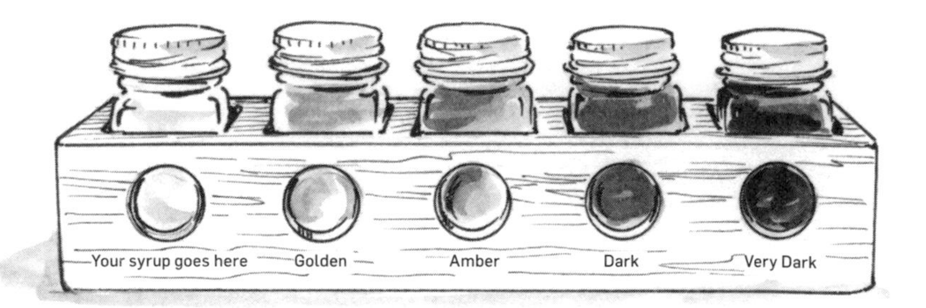


Test your syrup to see if it falls into one of four common pure maple syrup "grades." The fifth bottle in the comparator is for your syrup. Grade it against the four samples provided.

Maple producers have very little control over which grades of syrup they make. Typically, lighter syrup is made earlier in the season and darker syrup is made later. At the start of the season, your equipment is cleaner and the weather is colder. There are fewer bacteria contaminating your sap. Bacteria grow as the season continues and the weather warms, so toward the end of the season there are more bacteria contaminating your sap. These bacteria are not harmful — they are the same bacteria that are in the air we breathe. Bacteria metabolize the sugars in the sap, and this causes darker-colored, stronger-flavored syrup. The trees also go through chemical changes as they begin to bud, changes that affect the color and flavor of maple syrup. As with all aspects of syrup making, there are exceptions to these scenarios.

Dark syrup is not of inferior quality to light syrup, but different consumers prefer different grades of pure maple syrup. Just as with beer styles, color and taste differences in syrup grades do not denote differences in quality. A light pilsner beer has a light color and a delicate flavor. A stout beer has a dark color and a strong flavor. One style isn't of inferior quality to the other; they simply vary in color and taste.

## PACKAGING YOUR PRODUCT

MOST PURE MAPLE SYRUP that is packaged and sold on grocery store shelves is considered dark or amber grade. This is a fairly middle-of-the-road grade that will please the majority of consumers. When you're packaging for mass production, you have to keep this in mind. If you package smaller batches for individual customers, you can customize to their taste if you are lucky enough to have made the grade they prefer.

When packaging syrup for retail sale, IMSI suggests that it should be uniform in color; its flavor should fall into one of the four suggested classes; it should be free of foul or off-flavors; and it should be free of sediment and impurities. IMSI suggests that labels include the words "Grade A" and "Pure Maple Syrup," as well as where the product originated, the color class, the flavor intensity, and whether the grade is recommended for cooking. Additionally, labels could state that there are no additives. If your syrup is certified organic or kosher, add that to the label as well. Make sure your label has your name and at least the city and state in which your syrup is made so that customers can reorder from you when they run out.

## PRICING

To price your syrup for retail sale, first consider what price your syrup is worth in bulk. Packagers buy bulk syrup, usually in 30- or 55-gallon drums or barrels. Contact a packager to find out what price you would get if you sold your syrup to them. We're always happy to let people know the going rate for maple syrup. Our information is listed at the back of the book, and you can feel free to call. Prices may vary regionally, and they are not set for the year until after the syrup season is finished and bulk buyers are able to take inventory. Wait until the majority of the states and provinces have finished making syrup for the year before trying to pin down a price.

Start with that bulk syrup price, and increase from there, adding the cost of your containers and labels as well as other costs. Consider your time and the cost of your equipment. Consider labor: You're doing all the work of making the syrup, and you are also packaging, marketing, and storing it. Price it accordingly.

Also look at what syrup is selling for in the grocery store, and sell it for at least that price. People sometimes make the mistake of underpricing their syrup. Generally, when a customer goes to an individual producer to buy syrup, she does so because she feels that she is getting a higher-quality product. You don't need to price yourself out of the market, but making maple syrup is a very labor-intensive, specialized craft, and your prices should reflect that.

## SALE VENUES

IF YOU'RE BUILDING A BUSINESS, your business plan will need to show how your business will make a profit. You'll need to know what you're going to do with your syrup.

**Bulk.** Selling to a bulk buyer is the easiest way to sell your syrup. Ask other producers whom they sell to, or ask a maple syrup equipment dealer for names. If the dealer doesn't buy syrup, he or she will certainly know someone who does. Bulk buyers are almost always looking for more syrup and almost always willing to buy. Ask a trusted producer about the reputation of the buyer; you'll want to sell to a buyer who grades and prices syrup fairly.

**Retail.** Good places to start selling your syrup include farmers' markets, craft shows, and church bazaars. Some farmers' markets or bazaars have rules and fees, or they require a license to sell there. You'll want to make sure you contact the sale organizer before arriving. Some syrup makers find that they really enjoy selling their syrup this way; it gives them a chance to connect with each customer and talk about the process of making maple syrup.

Advertise that you have maple syrup to sell by putting a professional-looking sign at the end of your driveway.

It's always a good idea to put a sign at the end of your driveway; people won't know that you have syrup to sell if you don't advertise. Selling to your neighbors is a great way to start — they'll tell their neighbors, and word will spread that you make great maple syrup and you have some for sale!

**Wholesale.** Selling wholesale to gift stores, grocery stores, and even food service outlets is the next step. Try local or independent grocery stores first. It is difficult to gain access to chain stores, and they may require that you hire a broker and pay warehouse, shelf space, and other fees. Food service outlets such as bakeries and restaurants often cook or bake with bulk syrup or serve pure maple syrup. Try making connections with them as well.

## MARKETING YOUR PRODUCT

When marketing your pure maple syrup, start with the packaging. There are a variety of preprinted labels — ranging from simple to beautiful — that will work with any size or style of container. Some have space for you to add your business name and contact location. If they don't have space for that, we suggest sticking a return address label on the bottle. If you create your own label, make one that is unique to your business or family. It should clearly state "pure maple syrup."

**Proper language.** The maple syrup industry, with the help of IMSI and NAMSC, was instrumental in passing legislation stating that if the words "pure maple syrup" appear on a label there must be pure maple syrup in the container. There are severe penalties for breaking this law. Further, if the product is not pure maple syrup, the words "maple syrup" must be followed by a descriptor such as "flavored." For example, "maple syrup–flavored gum." This legislation was passed to protect consumers and the industry from fraudulent claims. If you are flavoring your pure maple syrup with, say, cinnamon, your label may state that it is "cinnamon-flavored pure maple syrup." If you are selling a blend of pure maple syrup and sugar syrup, which some producers do, you may not label your container as "pure maple syrup."

**Labeling grade.** Not all states require that you label the grade of your syrup, but it is important. In 2015 most states and provinces will require grade labels on pure maple syrup. We use grade labels so consumers know what they're buying. Grade labels also make your product look more professional. If they're having a label printed, people typically add the grade to the label design. If you decide to use a preprinted label, you can purchase grade-label stickers from a maple syrup supply dealer and place them on your container.

**State programs.** Many states offer specialty labels or programs. For example, Wisconsin offers the "Something Special from Wisconsin" program. Minnesota promotes the "Minnesota Grown" program. Each state has different program requirements, and many require a fee. These programs are beneficial because they have loyal followings. As the trend to buy local intensifies, consumers will continue to look for ways to identify local producers.

**Certification.** As consumers become savvier, they demand that more products be certified. There are many certifications — such as "organic," "kosher," and "gluten-free" — that maple syrup producers can use to describe their syrup. All of these certifications require yearly fees and inspections, tracking paperwork, and cleaning procedure and shipping documentation. For example, pure maple syrup is organic by nature, but you cannot label it organic without following the organic certification guidelines. Organic certification requirements pertain to land management, tapping procedures, the types of defoamers used during cooking, and products used during cleaning procedures. Contact the National Organic Program (see Resources for website) for more information on becoming certified organic.

For more information on how to market your pure maple syrup, check with the maple syrup producer association in your state or province. They may have suggestions or even marketing tools available for use.

A well-designed label can be an essential marketing tool for your syrup. Add any special certifications to your label. Your packaged syrup should also look professional. It should be wiped clean of any syrup drips, and the labels should be straight.

CHAPTER 11

# MAKING OTHER MAPLE PRODUCTS

Once you've mastered the art of making maple syrup, it is fun to try making maple sugar, candy, or cream. You can use special commercial candy machines and equipment, but they are not necessary for beginners — especially if you are making small amounts. With some practice you can make these products successfully on your stove top.

You don't add anything to the maple syrup to make these products. Sometimes pure maple cream is referred to as maple butter, but it does not actually contain butter or any other dairy product. Each product is made by continuing to evaporate water from the syrup. The amount of residual water that is evaporated, the way the boiling maple syrup is stirred or not stirred, and how quickly the syrup is cooled determines which type of product you make.

Making candy and cream is not a precise science, as the process can be greatly affected by the grade of syrup being used as well as the barometric pressure, air temperature, humidity, and elevation of your location. Experimentation and practice are essential parts of the process.

### Determine the Boiling Point of Water

*Remember that the boiling point of water varies depending on barometric pressure, so you will have to determine it prior to boiling your syrup. Do this every time you make a batch of maple sugar, candy, or cream. Bring water to a boil, and check the temperature with a candy thermometer.*

## MAPLE SUGAR

Maple sugar should be made with table grades of pure maple syrup (light, medium, or dark syrup). The coarseness of the finished sugar may vary depending on the grade of syrup that is used. For special equipment you will need a flat pan, a rolling pin, and a screen.

1. Boil pure maple syrup until it reaches a temperature that is between 34 and 38 degrees above the Fahrenheit boiling point of water (between 19 and 21 degrees above the Celsius boiling point; see box above). Watch the temperature carefully, as it can rise very quickly, and your syrup may burn.

2. When the syrup reaches the desired temperature, remove it from the heat, and stir until the syrup begins to crystallize and doesn't foam up any longer.
3. Pour the syrup onto a flat pan. If your syrup hardens before you get it out of your cooking pan, you can add water, reheat, and try again. Keep stirring the syrup and moving it around the flat pan while it cools. You will have granules and lumps of maple sugar. To remove the lumps, roll the sugar with a rolling pin and sift it through a screen.
4. Store your sugar in a cool, dry place.

Maple sugar is becoming a popular substitute for more traditional sweeteners. It can be used in baking, on hot or cold cereals, or in coffee.

## Smelling Smoke?

*If you smell smoke while you are cooking your syrup, remove your syrup from the heat. At this point, if the syrup hasn't already begun to burn, it will if you leave it over the heat any longer.*

## MAPLE CANDY

Maple candy can be made with light- to medium-grade pure maple syrup. Ideal candy will have a slight but consistent graininess throughout. To make candy you will need rubber candy molds.

1. Prepare your rubber candy molds following the manufacturer's instructions.
2. Boil pure maple syrup until it reaches a temperature between 32 and 34 degrees above the Fahrenheit boiling point of water (between 18 and 19 degrees above the Celsius boiling point of water; see box, page 107).
3. Remove the syrup from the heat, and without stirring cool the syrup to less than 200°F (93°C) but not lower than 160°F (71°C).
4. When it reaches the desired temperature, stir. As you stir, the syrup will become slightly thicker and lighter in color. Stir until the syrup becomes slightly creamy and opaque. This usually takes less than 5 minutes.
5. Pour the syrup directly into the candy molds. Cool your candy for 10 to 30 minutes, then remove it carefully from the molds. Store your candy in a cool, dry place.

Maple candy is a sweet treat that sells well at farmers' markets and roadside stands.

## MAPLE CREAM

Maple cream can be made with light- to medium-grade pure maple syrup. Good cream will not have any granules.

1. Prepare an ice water bath in your sink.
2. Boil pure maple syrup until it reaches a temperature between 22 and 24 degrees above the Fahrenheit boiling point of water (between 12 and 13 degrees above the Celsius boiling point of water; see box, page 107).
3. Remove the syrup from the heat, and place it in the ice bath sink. Without moving or stirring the syrup, cool it to 75°F (24°C). Stirring can cause your finished cream to become grainy. Once the syrup has cooled, it will be slightly firm to the touch.
4. Once it is cool, stir the cream slowly until it loses its glossiness and becomes opaque. When your cream is the consistency of smooth paste, you can transfer it to food-grade containers for storage and use.
5. Store your cream in the refrigerator. If the cream separates between uses, stir it until it returns to its original texture. Refrigerated maple cream will keep almost indefinitely, and frozen maple cream will probably keep forever!

Maple cream is becoming a popular spread for breads and muffins. It is harder to market for sale than maple sugar and candy because it needs to remain refrigerated.

# GLOSSARY

**Arch.** The part of an evaporator that produces heat, sometimes referred to as the firebox. The arch sits under the pans.

**Barometric pressure.** A measurement of atmospheric air pressure by a barometer.

**Baume.** A system of measurement that compares the density of maple syrup to that of a salt concentration of the same density. Sap has turned into syrup when it reaches 32 Baume.

**Brix.** A system of measurement that compares the density of maple syrup to that of a sugar solution with a known percentage of sugar. Sap has turned into syrup when it reaches 66 Brix.

**Ceramic blanket.** A high-temperature sheet insulation made from man-made vitreous fibers, used to line an evaporator arch.

**Collection area.** The open space in a drilled taphole once the spout is inserted. The collection area allows for sap to come to the surface through open wood grain. It also allows for sap to gather and create a small amount of pressure to push the sap through the spout.

**Defoamer.** A vegetable-based oil used to break the surface tension of boiling sap, thus reducing the amount of foam produced.

**Drop flue pan.** A pan used for cooking sap that has corrugated channels that drop down into the arch.

**Drop line.** Tubing that takes sap from the spout in the tree to the lateral line.

**Evaporator.** A unit used to boil maple sap; it includes a stainless steel pan or pans over a heat source called an arch. It was first made in the mid-1800s.

**Filter press.** A device with a series of metal plates lined with filter papers used to remove sugar sand and other impurities and particles from syrup.

**Firebrick.** A brick that will withstand high temperatures. It is used to line the inside of an evaporator arch.

**Flat pan.** A pan used for cooking sap that has a flat bottom.

**Float box.** A device that keeps the sap level in a cooking pan steady.

**Flue pan.** See **drop flue pan** and **raised flue pan**.

**Gravity flow.** A tubing system using natural gravity to draw sap to a collection point.

**Hooked spout.** A spout with a hook on the bottom side that holds a sap bucket.

**Hookless spout.** A spout without a hook on the bottom side. The sap bucket hangs from the spout itself and is kept in place by a notch on the top of the spout.

**Hydrometer.** See **sap hydrometer** and **syrup hydrometer.**

**Lateral line.** Tubing that runs no more than 100 feet through a woods. Lateral lines take sap from drop lines to mainlines.

**Mainline tubing.** Tubing that runs no more than 1,000 feet through a woods. Mainlines take sap from lateral lines to a holding tank.

**Mainline wire.** A high-tensile wire, either 9 or 12.5 gauge, that holds up the mainline.

**Orlon.** A wool-like, synthetic acrylic material developed by Dupont in 1941. The syrup industry uses Orlon to filter finished syrup and remove sugar sand.

**Paper prefilters.** Used with Orlon during the filtering process, paper prefilters remove larger unwanted particles from the syrup.

**Preheater.** A unit that preheats the sap before it enters the flue pan to prevent cold sap from coming in and slowing the evaporation process.

**Raised flue pan.** A pan used for cooking sap that has corrugated channels that rise up into the pan.

**Refractometer.** An instrument that determines the sugar content in syrup by measuring the refractive index of the syrup, or how light bends as it passes through the syrup.

**Releaser.** Part of a vacuum-flow tubing system, the device makes the connection and transference of sap and vacuum. It takes vacuum pressure from the vacuum pump and puts it into the sap lines while still drawing sap from the lines. It releases sap from the system into a collection tank.

**Reverse-osmosis machine.** A machine that uses pressure and a semipermeable membrane to separate water from the sugar particles in sap.

**Rule of 86.** An equation (86 divided by the sugar content of your syrup) that will tell you approximately how long it will take you to cook your sap, depending on its sugar content.

**Sap hydrometer.** A fragile glass instrument that measures the sugar content in sap.

**Sap sak holder.** An alternative to a sap bucket, it consists of a metal brace that holds a sap bag and hangs from a spout for the purpose of collecting tree sap.

**Side wire.** Wire that wraps around trees adjacent to the mainline. It is 14- or 16-gauge non-high-tensile wire and is used to keep the mainline straight.

**Smokestack.** Removes the smoke produced by fuel burning in the arch of an evaporator from the sugarhouse. It also creates a draft to draw air through the arch to help the fuel burn more efficiently.

**Spout.** A strawlike fitting made of metal or plastic that is inserted into a tree to allow sap to flow from the tree into a collection vessel. Other names for spouts include tap, spile, or spigot.

**Steam Away.** A preheater manufactured by Leader Evaporator that heats and evaporates water from sap before it enters the flue pan.

**Steam hood.** A unit that hangs over the top of the evaporator and directs steam out of the cookhouse.

**Steam stack.** A pipe that comes out the top of a steam hood that hangs over the boiling pans of an evaporator. Its function is to remove the steam produced by boiling sap from the sugarhouse.

**Sugar sand.** The grainy sediment that is created every time sap or syrup is boiled.

**Syrup hydrometer.** A fragile glass instrument that measures the sugar content in syrup.

**Tree scarring.** Vertical striations in a tree caused by a puncture. Sap does not flow freely through tree scarring, and you'll want to avoid it when tapping a tree.

**Tubing system.** A system of interconnected, semipliable tubes installed in a woods to collect maple sap.

**Vacuum flow.** A tubing system using a vacuum pump to draw sap to a collection point.

**Vacuum pump.** A pump that creates vacuum and transfers it into the lines of a tubing system.

**Wire ties.** Used to secure the mainline to the mainline wire.

# RESOURCES

We've offered a few options for collecting and cooking sap. Here are some additional resources. With practice and perseverance you can build a successful maple syrup business. Good luck!

## WHERE TO BUY SUPPLIES

**Anderson's Maple Syrup, Inc.**
715-822-8512
*www.andersonsmaplesyrup.com*

**Bascom Maple Farms**
603-835-6361
*www.bascommaple.com*

**Ben Meadows Company**
GHC Specialty Brands, LLC
800-241-6401
*www.benmeadows.com*
Tree-marking paint

**Leader Evaporator Co., Inc.**
802-868-5444
*www.leaderevaporator.com*

**The Nelson Paint Company**
800-236-9278
*www.nelsonpaint.com*
Tree-marking paint

**Sugar Bush Supply Co.**
517-349-5185
*www.sugarbushsupplies.com*

## INFORMATIONAL WEBSITES

**Anderson's Maple Syrup, Inc.**
*www.andersonsmaplesyrup.com*

**Cornell Sugar Maple Research & Extension Program**
*http://maple.dnr.cornell.edu*

**International Maple Syrup Institute**
*www.internationalmaplesyrup institute.com*

**Leader Evaporator Co., Inc.**
*www.leaderevaporator.com*

**National Organic Program**
Agricultural Marketing Service, USDA
*www.ams.usda.gov/nop*

**North American Maple Syrup Council**
*www.northamericanmaple.org*
Lists state and provincial associations

**Ohio State University Extension Maple Syrup Information**
*http://maplesyrup.osu.edu*

**Proctor Maple Research Center**
University of Vermont
*www.uvm.edu/~pmrc*

## BIBLIOGRAPHY

**Chapeskie, David John, and Lew J. Staats.** *Design, Installation and Maintenance of Plastic Tubing Systems for Sap Collection in Sugar Bushes: An Instruction Manual.* Eastern Ontario Model Forest, 2006.

**Heiligmann, Randall Bruce, Melvin R. Koelling, and Timothy D. Perkins, eds.** *North American Maple Syrup Producers Manual,* 2nd ed. Ohio State University Extension, 2006.

**Perrin, Noel.** *Making Maple Syrup: The Old-Fashioned Way.* A Storey Country Wisdom Bulletin, A-51. Storey Publishing, 1980.

### Metric Conversion Chart

| WHEN THE MEASUREMENT GIVEN IS | TO CONVERT IT TO | MULTIPLY IT BY |
|---|---|---|
| inches | centimeters | 2.54 |
| feet | meters | 0.305 |
| square feet | square meters | 0.093 |
| ounces | grams | 31.1 |
| pounds | kilograms | 0.373 |
| tons | metric tons | 0.907 |
| gallons | liters | 3.785 |
| gallons | imperial gallons | 0.83 |

# INDEX

Page numbers in *italic* indicate illustrations; those in **bold** indicate charts.

# T



# W